Principles of
Hydraulic System Design

First Edition

Coxmoor Publishing Company's

Principles of Hydraulic System Design

First Edition

Author: Peter Chapple

ISBN 1 901892 15 8

Copyright 2003 © COXMOOR PUBLISHING COMPANY
Principles of Hydraulic Systems Design
First edition 2003
All rights reserved

This book is sold subject to the condition that it shall not by way of trade or otherwise be resold, lent, hired out, stored in a retrieval system, reproduced or translated into a machine language,or otherwise circulated in any form of binding or cover other than that in which it is published, without the Publisher's prior consent and without a similar condition including this condition being imposed on the subsequent purchaser.

Other titles published by the Coxmoor Publishing Company include:

Introductory Texts

*Vibration
Wear Debris Analysis
Thermography
Corrosion
Appearance & Odour
Oil Analysis
Acoustic Emission & Ultrasonics
Level, Leakage & Flow
Load, Force, Strain & Pressure
Power, Performance, Efficiency & Speed
Temperature
Condition Monitoring – An Introduction & Multi-lingual Dictionary*

Conference Proceedings

Condition Monitoring '99
COMADEM '99
Condition Monitoring 2001

Published by
Coxmoor Publishing Company
PO Box 72, Chipping Norton,
Oxford OX7 6UP, UK
Tel: +44 (0) 1451 - 830261
Fax: +44 (0) 1451 - 870661
E-mail: mail@coxmoor.com

Printed in Great Britain by Information Press, Oxford, UK

THE BRITISH FLUID POWER ASSOCIATION

The British Fluid Power Association represents the interests of over 85% of the British manufacturers and suppliers of hydraulic and pneumatic components and systems, with sales of over £800 million.

The BFPA offers a wide range of services to its member companies, including:

- Technical Standards and Guidelines - monitoring legislation, co-ordinating research projects, liaison with BSI, CEN and ISO, formulating industry proposals and views on technical issues, producing guidelines and codes of practice.

- Education and Training - developing courses and qualifications, liaising with educational and training authorities, universities and colleges.

- Market and Statistical Data - collecting industry statistics, import and export information, market forecasting.

- Promotion and Marketing - publishing newsletters, directories, websites, industry and product information.

- Exhibitions and Conferences - organising Motion and Control 2004 (incorporating IFPEX), attending overseas exhibitions, arranging group stands.

- Representation of Industry's views - through contact with Government departments, procurement agencies, overseas consulates, the media.

- Providing a meeting place within the Industry - discussing matters of common concern and interest.

- International co-operation - through CETOP and fluid power trade associations worldwide.

For further information please contact:
Mr Roman Russocki, Chief Executive, British Fluid Power Association
Cheriton House, Cromwell Park, Chipping Norton, Oxon OX7 5SR
Tel: + 44 (0)1608 647900   **Fax:** + 44 (0)1608 647919
E-mail: enquiries@bfpa.co.uk   **Website:** http://www.bfpa.co.uk

BFPA — The British Fluid Power Association

Total Power & Control...

...from a single source

Hydraulic Pumps, Motors, Valves and Systems for all applications...

Denison Hydraulics, the innovative manufacturers of pumps, motors, valves, electronic controls and systems, offers a full product range for every industry.

One supplier - TOTAL SUPPORT.

Product Quality & Reliability is the key to success in all industries, who take the advantage of many product features.

- Smooth and precise position control
- Lightweight, compact and efficient systems
- Fast and reliable controlled response to pressure surges
- Full proportional control with accurate positioning and repeatability

The Innovation and the Answers For full control of your systems...

...contact

DENISON Hydraulics

Kenmore Road, Wakefield 41 Industrial Estate, Wakefield, West Yorkshire, WF2 OXE, England.
Tel. +44 (0)1924 826021 Fax. +44 (0)1924 826146

Foreword

This book has been prepared to provide the basis for the BFPA course 'Principles of Hydraulic Fluid Power System Design' under the auspices of the BFPA Education and Training Committee.

Fluid power systems are manufactured by many organisations for a very wide range of applications, which often embody differing arrangements of components to fulfil a given task. Hydraulic components are manufactured to provide the control functions required for the operation of systems, each manufacturer using different approaches in the design of components of any given type.

As a consequence, the resulting proliferation of both components and systems can, to the uninitiated, be an obstacle to the understanding of their principle of operation. The book is structured so as to give an understanding of:

- The basic types of components and their operational principles.
- The way in which circuits can be arranged using available components to provide a range of functional outputs.
- The analytical methods that are used in system design and performance prediction.

Components are arranged to provide various generic circuits, which can be used in the design of systems so as to suit the functional characteristics of the particular application. In order to aid the understanding of the mechanical principles involved, diagrams of some manufacturer's components are included in the book. These are shown to give examples only and are not comprehensive, and have been reproduced from technical literature. The author wishes to acknowledge the availability of such material.

P J Chapple
30th September 2002.

Flexible Control

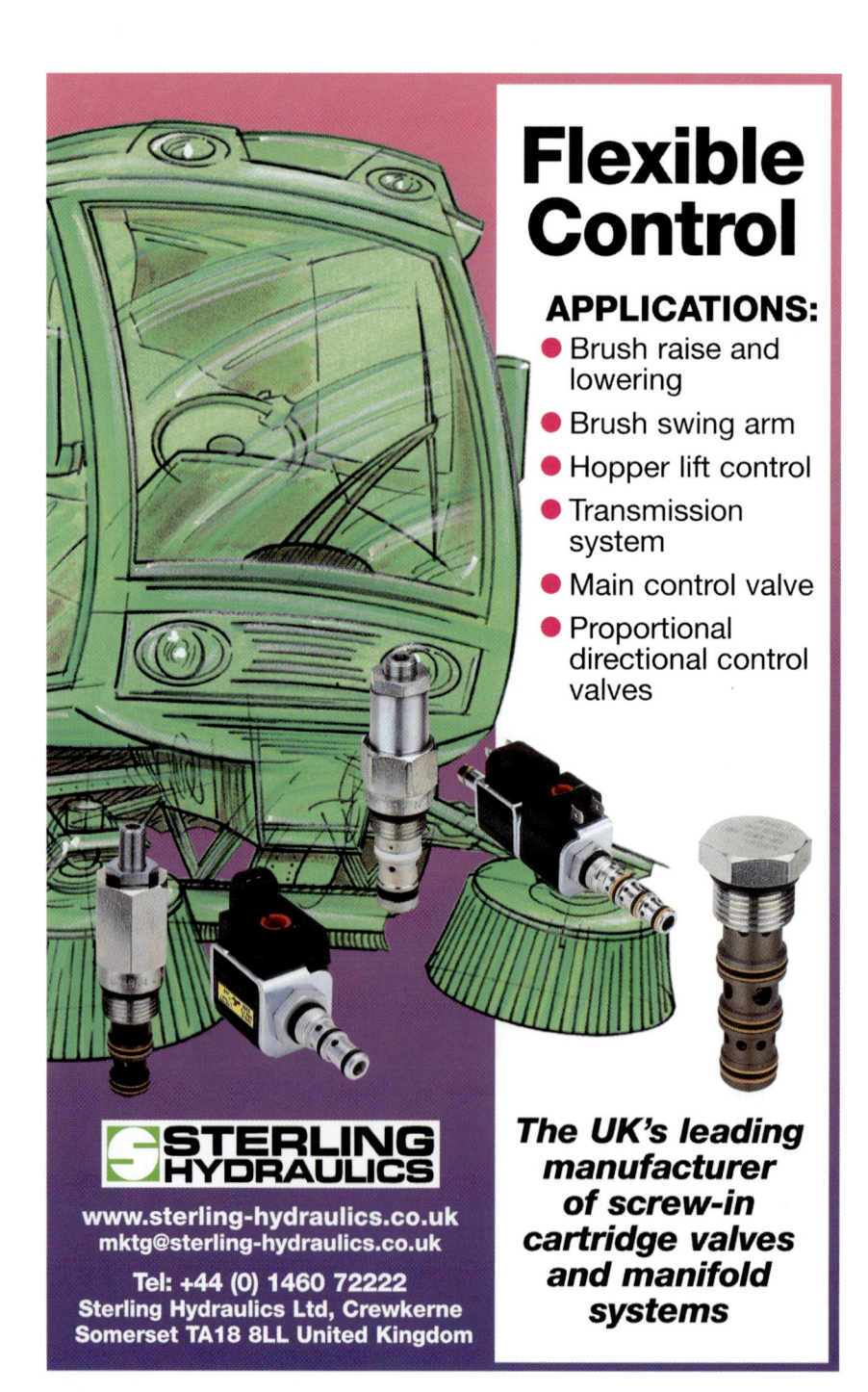

APPLICATIONS:
- Brush raise and lowering
- Brush swing arm
- Hopper lift control
- Transmission system
- Main control valve
- Proportional directional control valves

STERLING HYDRAULICS

www.sterling-hydraulics.co.uk
mktg@sterling-hydraulics.co.uk

Tel: +44 (0) 1460 72222
Sterling Hydraulics Ltd, Crewkerne
Somerset TA18 8LL United Kingdom

The UK's leading manufacturer of screw-in cartridge valves and manifold systems

Contents

Graphic Symbols For Fluid Power Systems xxi

Chapter One
HYDRAULIC POWER TRANSMISSION
1   Introduction 3
2   Fluid power system design 4
3   Contents of the book 5

Chapter Two
HYDROSTATIC PUMPS AND MOTORS
       Summary 9
1   Introduction 9
2   Major aspects in the selection of pumps and motors 10
3   Types of pumps and motors 11
3.1   Fixed displacement units 11
3.1.1   External gear pumps/motors 11
3.1.2   Internal gear pumps 12
3.1.3   Vane pumps/motors 13
4   Variable displacement units 14
4.1   Vane pumps 14
4.2   Piston pumps/motors 14
5   Equations for pumps and motorss 16
5.1   Flow and speed relationship 16
5.1.1   Volumetric efficiency 17
5.2   Torque and pressure relationship 17
5.2.1   Mechanical efficiency 17
5.3   Pump selection parameters 19
6   Low speed motors 20
6.1   Types of low speed motors 20
6.1.1   Radial piston motors 20
7   Some general considerations 22
8   Comparison of Motor Performance Characteristics 23

Chapter Three
HYDRAULIC ACTUATORS
       Summary 29
1   Introduction 29
2   Linear actuators 30
3   Principle features 30
3.1   End covers 30
3.2   Mounting methods 30
3.3   Seals 33
3.4   Position transducers 33

Linde
We move the world

Hydraulics product programme

Linde's axial-piston units lead the European market for construction and agricultural equipment, and the Linde hydrostatic transmission has established itself worldwide as the optimum drive system for all mobile equipment.

United Kingdom
Linde Hydraulics Ltd
7 Nuffield Way
Abingdon, Oxfordshire OX14 1RJ
Tel ++44 (0) 1235 522828   Fax ++44 (0) 1235 523184
E-Mail: enquiries@lindehydraulics.co.uk   www.lindehydraulics.co.uk

4	Actuator selection	33
4.1	Actuator force	33
4.2	Cushioning	34
5	Rotary actuators	37
5.1	Actuator types and capacity range	37
5.2	Applications	38

Chapter Four
PRESSURE CONTROL VALVES

	Summary	41
1	Introduction	41
2	Relief valves	42
2.1	Single stage relief valve	42
2.2	Two Stage Relief Valves	43
3	Pressure Reducing Valves	44
4	Counterbalance Valves	46

Chapter Five
FLOW CONTROL VALVES

	Summary	53
1	Introduction	53
2	Directional control valve	54
3	Restrictor valve	56
4	Pressure compensated valve	56
5	Central bypass valve	56

Chapter Six
ANCILLARY EQUIPMENT

	Summary	63
1	Introduction	63
2	Accumulators	64
2.1	Types	64
2.2	Performance	65
3	Contamination control	67
3.1	Components	67
3.2	Filters	68
4	Coolers	72
4.1	Cooler types	72
4.2	Thermodynamic aspects	73
4.3	Cooler characteristics	73
5	Reservoirs	73

Chapter Seven
CIRCUIT DESIGN

	Summary	77

the solution is in the media

Introducing **Pall Ultipor SRT** filter elements

The Next Generation in Fluid System Protection featuring **Stress Resistant Media Technology**

Ultipor SRT

Designed and manufactured to be resistant to hydraulic system stresses

- Assured performance under cyclic conditions
- Tested and approved to the new Stress Resistance Test
- Increased flow capacity
- Smaller package size
- **Reduced filtration costs**

Ultipor SRT Filtration

Ultipor SRT
...a filter for modern day demands

PALL Pall Corporation

For more information on Ultipor SRT media technology, contact Pall on:

+44 (0)23 9230 3303 tel
+44 (0)23 9230 2507 fax
m&e_sales@pall.com

1	Introduction	77
2	Pressure and Flow	78
3	Directional control	79
3.1	Two position valves	79
3.2	Three position valves	80
4	Load holding valves	82
5	Velocity control	83
5.1	Meter-in control	83
5.2	Meter-out control	85
5.3	Bleed-off control	86
5.4	Four way valve restrictive control	87
5.4.1	Analysis of the valve/actuator system	88
5.4.1.1	Actuator extending	88
5.4.1.2	Actuator retracting	91
5.4.2	Valve sizing	92
5.4.3	Valves with non-symmetrical metering	94
5.5	Bypass control with fixed displacement pumps	95
5.5.1	Open centre valves	95
5.5.2	Closed centre valves with load sensing and pressure compensation	97
6	Variable displacement pump control	98
6.1	Load sensing	99
6.2	Power control	99
6.3	Accumulator charging	101
7	Hydrostatic transmissions	102
7.1	Pump controlled systems (primary control)	102
7.2	Motor brake circuit	103
7.3	Linear actuator transmissions	103
7.4	Motor controlled systems (secondary control)	104
8	Pilot operated valve circuits	105
8.1	Load control valves	105
8.2	Pump unloading circuit	105
8.3	Sequence Control	106
9	Contamination control	106

Chapter Eight
FLOW PROCESSES IN HYDRAULIC SYSTEMS

	Summary	111
1	Introduction	112
2	Fluid properties	112
2.1	Fluid viscosity	112
3	Flow in pipes	113
4	Laminar flow in parallel leakage spaces	116
5	Orifice flow	117

WEBTEC
WEBTEC PRODUCTS LIMITED

Manufacturers of hydraulic components and test equipment for the mobile, industrial and agricultural industries

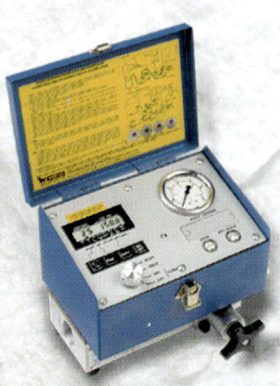

The Webster range of hydraulic test equipment

- Portable Testers
- Flow Indicators 
- Pressure Testing 
- Data Acquisition Systems
- Flow meters and Readouts
- Contamination monitors

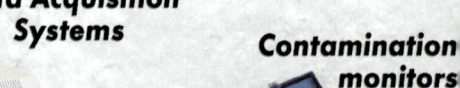

Nuffield Road, St. Ives, Cambridgeshire, PE27 3LZ, UK
www.webtec.co.uk   Tel: 01480 397400
E-mail: sales@webtec.co.uk   Fax: 01480 466555

6	Valve force analysis	118
6.1	Poppet valves	119
6.1.1	Momentum force	119
6.1.2	Valve flow	120
6.1.3	Valve pressure/flow characteristics	120
6.2	Spool valves	121
	Appendix	123

Chapter Nine
OPERATING EFFICIENCIES OF PUMPS AND MOTORS

	Summary	127
1	Introduction	127
2	Mechanical and volumetric efficiency	127
3	Analysis of the losses	128
3.1	Theoretical performance	128
3.2	Volumetric flow loss	129
4	Mechanical loss	130
5	Unit efficiency	131
5.1	Volumetric efficiency	131
5.2	Mechanical efficiency	131
5.3	Overall efficiency	131

Chapter Ten
CONTROL SYSTEM DESIGN

	Summary	137
1	Introduction	138
2	Simple valve actuator control	138
2.1	Open loop system	138
2.2	Closed loop system	140
2.3	System response	142
3	Fluid compressibility	143
3.1	Bulk modulus	143
3.2	Hydraulic stiffness	143
4	Valve actuator dynamic response including compressibility effects	144
4.1	Valve flow	144
4.2	Actuator flows	148
4.3	Actuator force	149
4.4	Comments	150
4.5	Actuator position	150
4.6	Valve selection	151
4.7	Pressure shock control in open loop systems	151
5	Frequency response	151
5.1	Simple actuator	152
5.2	Valve actuator system	153

Pipe Clamps

Testing

Filtration

STAUFF®
www.stauff.com

The best in hydraulic pipework
- components
- on site design
- prototyping
- sub assemblies
- kitted packages
- managed supply systems

In partnership with

WALTERSCHEID

Accessories

Tube Connectors and tube forming

Hose assemblies

With full MOD approval for Walform tube connector systems, you can share in the benefits of a genuine leak free system without incurring any price premium

For year round peace of mind, national or local supply and global support

Stauff UK
332, Coleford Road, Darnall, Sheffield S9 5PH
Tel: 01142 518 518   Fax: 01142 518 519

6	Stability of the closed loop position control system	154
6.1	Stability criterion	154
6.2	System design	154
6.3	Steady state accuracy	156
6.3.1	Valve leakage	157
6.3.2	Valve hysteresis	160
7	The improvement of closed loop system performance	161
7.1	Position control	161
7.2	Velocity control	161
7.3	Pressure control	162
8	Compensation techniques	164
8.1	Integral plus proportional compensation	164
8.2	Proportional plus derivative control	165
8.3	Phase advance compensation	166
8.4	Proportional, Integral and Derivative (PID) control	167
8.5	Pressure feedback	168
9	System frequency response tests	169
10	Pump controlled systems	170

Chapter Eleven
PERFORMANCE ANALYSIS

1	Introduction	175
2	Meter-in control	175
2.1	The effect of load force changes	177
3	Valve control of a single ended actuator	177
3.1	Data	177
3.2	Actuator retraction	178
3.3	Actuator extension	179
4	Winch Application	179
4.1	Lifting the load	180
4.2	Lowering the load	180
4.3	Numerical Values	180
5	Hydraulic Motor for Driving a Winch	181
5.1	Gearbox ratio	182
5.2	Motor selection	183
5.3	Flow required	183
5.4	Minimum motor displacement	183
6	Hydraulic system for gantry crane	183
6.1	Gantry crane	185
6.2	Wheel drive	187
6.3	Pipe sizes	188
7	Pressure losses	190
7.1	Data	190
7.2	Pressure loss at 20^0C	191
7.3	Pressure loss at 60^0C	192

UNDER
PRESSURE?

Hydraulic screw-in cartridge valves and manifold blocks

- Reliable
- Efficient
- Cost effective
- Compact
- Serviceable
- Versatile

As one of the world leaders in the design & manufacture of hydraulic screw-in cartridge valves we offer:

- Improved Performance
- Reduced Costs
- Peace of mind

Let us take the pressure off you and put it back where it belongs!

Call: +44(0)1926 881171 or (440) 974 3171

ih Integrated Hydraulics

Integrated Hydraulics Limited
Collins Road, Heathcote Industrial Estate, Warwick, CV34 6TF, United Kingdom
e-mail. ukinfo@integratedhydraulics.co.uk

Integrated Hydraulics Inc
7047 Spinach Drive, Mentor, Ohio 44060-4959, USA.
e-mail: usinfo@integratedhydraulics.com

7.4	Pump requirements	192
8	Single stage relief valve	193
9	Simple actuator cushion	194
10	Central bypass valve	196
10.1	Flow analysis	197
10.2	Valve characteristics	199
10.3	Actuator force	200
10.4	Valve operation	201
11	Pump and motor efficiencies	202
12	Control System Design	204
12.1	Data	204
12.2	System gain	205
13	Hydraulic system for injection moulding machine	207
13.1	System data	207
13.2	Injection moulding machine	207
13.3	Actuator	207
13.4	Motor	208
13.5	Volume required/cycle	208
13.6	Accumulator	208
13.7	Circuit	209
13.8	The prevention of pressure shocks	210
14	Oil Cooling	211
14.1	System data	211
14.2	Duty cycle	212
14.3	Heat generated	212
14.4	Heat loss to the surroundings	215
14.5	Pump efficiency	215
15	Vehicle Transmission	215
16	Pump Control Applications	222
16.1	Application of Pump Unloader Valve to Vehicle Crusher/Refuse Machines	222
16.2	Application of Pump Controls to Bending Machines	223
17	Application of compensation techniques	225
17.1	Steady state accuracy	225
17.2	Proportional plus integral compensation	226
17.3	Proportional plus integral plus derivative (PID) compensation	227
17.4	Load pressure feedback	228
17.5	Concluding remarks	229

Chapter Twelve
SYSTEMS MANAGEMENT

	Summary	233
1	Introduction	233
2	Aspects of systems management	234
3	Systems management objectives	234

4	System cleanliness	235
5	Fault analysis	236
5.1	Fault Tree Analysis (FTA)	236
5.2	Failure modes effects analysis (FMEA)	237
5.2	Failure modes effects analysis (FMEA)	237
6	General	238

INDEX 241

Graphic Symbols For Fluid Power Systems

Basic Symbols

——————	Working Line - Pressure/Return
— — — —	Pilot Control Line - External/Internal Drain Line
— - — - —	Enclosure of two or more functions contained in one unit
⌒	Flexible Connection
┬ ┴	Pipe Line Junction
+ ⤴	Crossing - Not Connected
▼	Hydraulic Source of Energy
▽	Pneumatic Source of Energy
↑↓↑	Directions of Flow
((	Directions of Rotation
↗	Facility for Adjustment of: - Spring, Pump, Solenoid, Cylinder, Damping, etc
⟋⟍⟋	Spring
≍	Restriction to Flow Line
——✕——	Power Line Plugged
⊢○⊣○⊢	Quick Release - Self Sealing Coupling - Connected (Mechanically Opened)
⊢○⊣	Coupling - Blocked e.g. Gauge Point
⊖	Rotary Junction - One Connection
≡⊖≡	Rotary Junction - Three Connections
═╪═	Rotary Shaft - Uni-directional
═╪═	Rotary Shaft - Bi-directional
═╤═	Shaft - Detent Held
⟊	Pivoting Device

Pumps

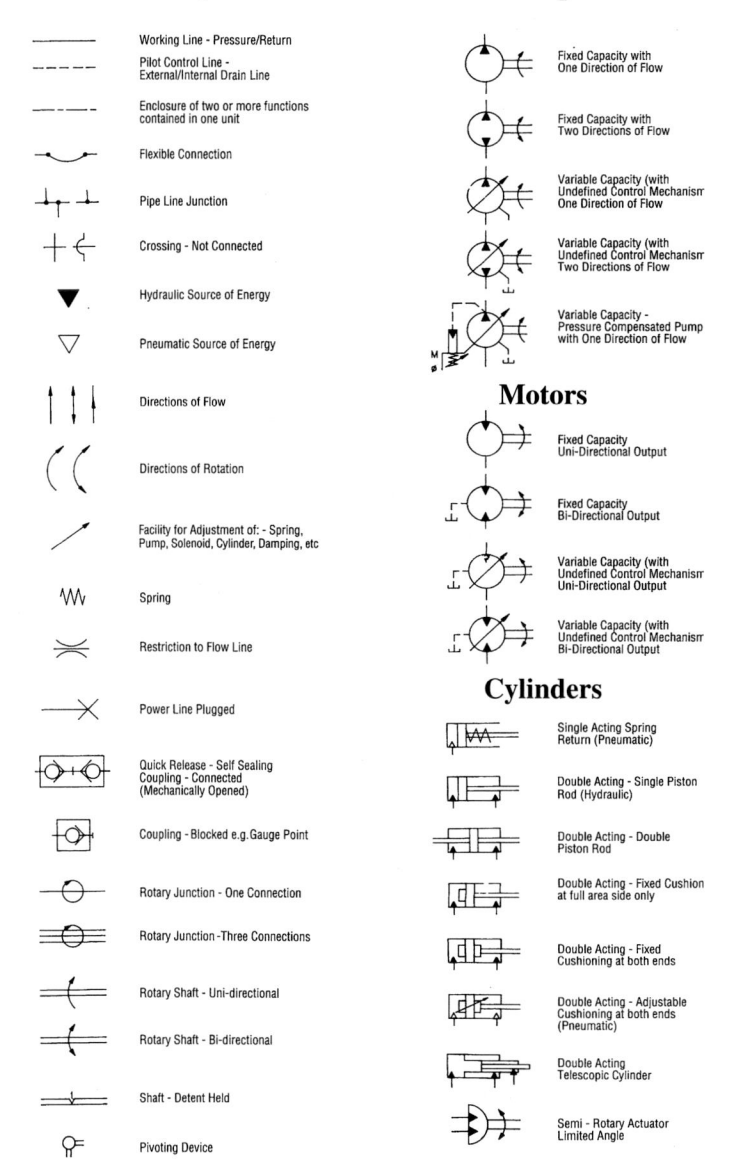

- Fixed Capacity with One Direction of Flow
- Fixed Capacity with Two Directions of Flow
- Variable Capacity (with Undefined Control Mechanism) One Direction of Flow
- Variable Capacity (with Undefined Control Mechanism) Two Directions of Flow
- Variable Capacity - Pressure Compensated Pump with One Direction of Flow

Motors

- Fixed Capacity Uni-Directional Output
- Fixed Capacity Bi-Directional Output
- Variable Capacity (with Undefined Control Mechanism) Uni-Directional Output
- Variable Capacity (with Undefined Control Mechanism) Bi-Directional Output

Cylinders

- Single Acting Spring Return (Pneumatic)
- Double Acting - Single Piston Rod (Hydraulic)
- Double Acting - Double Piston Rod
- Double Acting - Fixed Cushion at full area side only
- Double Acting - Fixed Cushioning at both ends
- Double Acting - Adjustable Cushioning at both ends (Pneumatic)
- Double Acting Telescopic Cylinder
- Semi - Rotary Actuator Limited Angle

Principles of Hydraulic System Design

Control Valves

Single Stage - Relief Valve (Adjustable)

Two Stage Relief Valve - Internal/External Pilot operated (Adjustable)

Two Stage Sequence Valve - Internal/External Pilot operated (Normally Closed: Adjustable)

Two Stage Proportional Pilot Operated Relief Valve

Single Stage Reducing Valve (Adjustable)

Two Stage Reducing/Relieving Valve - Internally Pilot Operated (Adjustable)

Gate Valve

Check Valve - Spring Closed

Pilot to Open Check Valve

Adjustable Restrictor Valve

Variable Pressure Compensated Flow Control Valve

Priority Flow Control

Flow Divider / Combiner Fixed Ratio Outputs

One Way Adjustable Restrictor Valve - With Free Flow in One Direction & Restricted Flow in the Other Direction

Shuttle Valve

Quick Exhaust Valve (Pneumatic)

Controls

By Twist

By Push Button

By Pull Button

By Lever

By Foot Pedal

By Plunger

By Roller

By Spring

By Single Solenoid (One Winding)

By Single Solenoid (Two Opposed Windings)

By Electric Motor

By Pneumatic Pilot Pressure (Direct)

By Differential Pilot Areas

By Oil Pressure (via Control)

By Pneumatic Pressure (via Control)

By Internal Control Path

By External Control Path

By Solenoid Operated Pilot Valve

Note:
Valve control symbols may be drawn at any convenient position across end of valve rectangle

Principles of Hydraulic System Design xxiii

Directional Control Valves

	Two Position Valve - Two Port Flow Paths Blocked/Through
	Two Position Valve - Four Port Flow Paths Through/Crossed
	Two Position - Two Port Spring Return Pilot Operated Valve (Pneumatic)
	Two Position Two Port Push/Pull Detent Held Valve
	Two Position Three Port Single Solenoid Operated Spring Return Directional Valve (Transient Port Connection Shown)
	Two Position Four Port Solenoid Pilot Operated Directional Valve with Spring Return (External Solenoid Pilot & Drain Connection - Solenoid with Mechanical Override)
	Two Stage Five Port - Solenoid Pilot Operated Pneumatic Directional Valve
	Electro Hydraulic Servo Valve Spring Centred with Variable Flow Area - Four Port Type Single Actuator - Two Windings
	Proportional Solenoid Pilot Operated Directional Valve with Variable Flow Area Four Port - Two Actuators
	Logic Valve - Pilot Operated, Spring Biased with Differential Pilot Areas (Normally Closed Spool Type)
	Logic Valve - Pilot Operated, Spring Biased with Differential Pilot Areas (Normally Open-Seated Spool)

Miscellaneous

	Air Breather
M	Electric Motor
	Reservoir - Return Above Oil Level
	Reservoir - Return Below Oil Level
	Accumulator - Fluid Maintained Under Pressure by Inert Gas
	Filter
	Pneumatic Silencer
	Filter with Water Trap (Manual Drain)
	Filter with Water Trap (Automatic Drain)
	Lubricator
	Cooler (No Indication of Cooling Method)
	Pressure Gauge
	Pressure Indicator
	Heater
	Level Gauge
	Thermometer
	Flow Meter
	Pressure Switch (Hydraulic)
	Thermostat
	Oil Level Switch (Two Level Switching)
	Intensifier - Air / Oil

CHAPTER ONE

HYDRAULIC POWER TRANSMISSION

1. HYDRAULIC POWER TRANSMISSION

1. Introduction

Hydraulic fluid power is one of the oldest forms of power transmission which, despite the period of rapid growth of electric power generation, became accepted for driving a wide range of machines because of the inherent advantages that it has over other available forms of power transmission.

The increased reliability and life that resulted from the introduction of oil based fluids and nitrile rubber sealing elements created a rapid growth in the use of fluid power transmission systems for a large variety of machine applications. Some of the advantages that hydraulic power has over other transmission mediums are summarised:

- The equipment designer is released from the dimensional limitations that are imposed by conventional gears and driveshafts.
- Stepless speed control can be obtained with relatively little increase in circuit complexity.
- The high ratio of power to mass allows fast response and a low installed weight at the point of application.
- The available output force is independent of operating speed. Stalled loads can be maintained for indefinite periods.

The introduction of electronic control into fluid power has created scope for its use in a wide range of machine applications particularly when operation by computers or PLCs is required. Electronic devices have improved the accuracy of control using closed loop control techniques in many applications that have traditionally been served by hydromechanical open loop systems.

Whether or not fluid power transmission is adopted in a particular application depends on a number of features that would require consideration for a comparative study to be made if different types of power transmission are to be evaluated.

2. Fluid power system design

There are broad categories of the types of fluid power systems in normal use for which there is a range of available components for any chosen system. The type of circuit employed often depends on company practice or user choice and, as a consequence, this often has an important influence on the components selected for the system. However, there are technical aspects that can be used to evaluate the performance of systems, which the designer needs to be aware of in order to provide some influence in the process of selecting both the type of circuit to be used and the components.

2.1 Component selection

Circuits can be arranged in various ways using alternative components to provide a system for any given application. Additionally, different component designs are available to perform a specific function and because of this and their influence on circuit design, the component selection process does not easily lend itself to a discrete synthesised approach, as it requires knowledge of the:

- Range of hydraulic components that are available.
- Operating characteristics of the components and their use in circuits and control systems.
- Available types of hydraulic circuits.
- Analytical methods for determining the system performance to meet the machine specification.

2.2 Circuit selection

Generally speaking the type of circuit that is chosen for a given application depends on a number of factors that include:

- First cost
- Weight
- Ease of maintenance
- Operating cost
- Machine duty cycle

2.3 System design process

The major activities involved in the design process can be summarised as follows:

- Evaluate the machine specification and determine the type of hydraulic system to be used.
- Establish the types and sizes of the major hydraulic components.
- Select an appropriate design of the hydraulic circuit.
- Carry out a performance analysis of the system and determine its ability to meet the machine specification.

This process, or parts of it, may need to be repeated as the final design is evolved.

3. Contents of the book

Bearing in mind the foregoing comments, this book has been arranged to provide background knowledge for the design of fluid power systems. The contents include:

- Descriptions of major hydraulic components and circuits and their performance characteristics.
- Methods for analysing the flow in pipes and components and flow forces on valves.
- The modelling of the efficiency of pumps and motors.
- Techniques for the design and analysis of control systems.
- Methods for the analysis of system performance.

This book is, therefore, aimed at providing the required background knowledge in the design of hydraulic fluid power systems and their application to a wide range of engineering equipment and machines.

CHAPTER TWO

HYDROSTATIC PUMPS AND MOTORS

2. HYDROSTATIC PUMPS AND MOTORS

Summary

There is a wide range of available hydrostatic pumps and motors from the market and the purpose of this chapter is to describe the operating principles and features of the most commonly used types. The formulae that are used for determining the performance of pumps and motors are presented and some of the major parameters that can be used as a basis for comparison are outlined as a background for the selection process. However, because of the wide variety of the types of units that are available it is impossible to generalise on the selection process in any given application.

Commonly machine builders and users have preferences for particular types of pumps and motors that are based on experience with particular applications which are determined by factors such as the system function, its control, servicing aspects, environmental features, life expectancy, duty cycle and type of fluid to be used. The designer needs to be aware of the relative performance of the different types and how this knowledge can be utilised in the selection process to suit a particular application.

1. Introduction

Power transmission pumps in fluid power systems are usually hydrostatic or positive displacement units, which convert mechanical power into fluid power, the most common types being gear pumps, vane pumps and piston pumps. In these pumps fluid is transferred through the machine in discrete volumes e.g. a gear tooth cavity. The pump size and speed determines the fluid flow rate.

Hydrostatic pumps are sources of flow so that when they are connected to a hydraulic motor, the outlet pressure will rise so that the flow can cause the motor to rotate against the load torque. Hydrostatic motors convert fluid power into mechanical power so that rotation of the output shaft can take place against an opposing torque load. Generally speaking pumps can be run as motors but a number of factors influence this possibility, some of which are:

- Not all pumps are reversible in direction of rotation because of their internal and external sealing arrangements.

- Pumps are designed to operate at relatively high speeds and can be inefficient at low speeds particularly during starting.

- Motor applications often require a significant shaft side load capacity. Pump rotating components are generally not designed to carry such shaft side loads and consequently cannot be directly coupled to the output drive where side loading exists.

This chapter is concerned with describing the operating principles of hydrostatic units, some aspects involved in their selection and the determination and presentation of their performance characteristics.

2. Major aspects in the selection of pumps and motors

The selection of pumps can be determined by a number of factors, which need to be considered by the user. These factors include:

- Cost.
- Pressure ripple and noise.
- Suction performance.
- Contaminant sensitivity.
- Speed.
- Weight.
- Fixed or variable displacement.
- Maximum pressure and flow, or power.
- Fluid type.

3. Types of pumps and motors

The mechanical principle that is chosen in the design of high-pressure positive displacement pumps and motors, which includes those using pistons, vanes and various gear arrangements depends on a number of factors. These include the operating speed and pressure, the type of fluid and the requirement for providing variable displacement control.

Pumps normally operate at constant speed (e.g. driven by electric motor) although in some situations (e.g. those driven by an internal combustion engine as found typically in mobile applications) the speed will vary over a small range. However, for motors it is normally required to operate at varying speeds including starting from rest (e.g. winch drives) and this aspect is reflected in the design of some available types.

Positive displacement machines are quite distinct from those using rotodynamic principles, which are often used for the transfer of fluid at relatively high flow rate and low pressures. Positive displacement units operate at relatively low flow rates and high-pressure and normally can only be used with fluids having good lubricating properties. However, there are machines that can be used with fire resistant fluids and pure water.

3.1 Fixed displacement units

3.1.1 External gear pumps/motors

In many applications, for operation at pressures up to 250 bar, external gear pumps/motors are used extensively because of their simplicity, low cost, good suction performance, low contamination sensitivity and relatively low weight. In applications requiring low noise, vane or internal gear pumps are often used.

Essentially the unit consists of two meshing gear pinions, mounted in bearings and contained in a housing or body as shown in Figure 1. As the pinions are rotated, oil is trapped in the spaces between the gear teeth and the housing and carried round from the pump inlet to its outlet port when the trapped volume is discharged by the action of the gears meshing together.

Torque is required at the input shaft at a level that is dependent on the outlet pressure acting on the gear teeth. When supplied with high-pressure flow the unit acts as a motor by providing torque to drive the load on the output shaft.

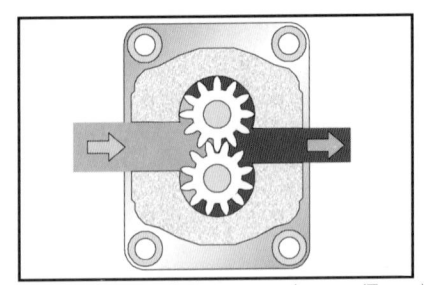

Figure 1. External Gear pump/motor (Eaton).

Some of the outlet fluid is transferred back to the low pressure side by way of small leakage flows through the:

- Clearance space between the teeth and the housing.
- Shaft bearing clearances.
- Clearance between the gear faces and the side plates in the housing. Most gear units have pressure loaded side plates to minimise this leakage.

The design of the unit is such as to minimise flow losses as they reduce its efficiency, particularly when using fluids of low viscosity such as with some water based fluids. The geometric capacity, or displacement, cannot be varied so their displacement is fixed. For a given gear form the manufacturer can produce pumps of different displacements by using different gear widths.

Standard types operate at speeds of 1000 to 3000 RPM and at pressures up to 250 bar but higher speeds and pressures are available. Powers range from 1 to over 100 kW. The efficiency of gear units has been raised during recent years, with peak overall efficiencies of 90% or above.

3.1.2 Internal gear pumps

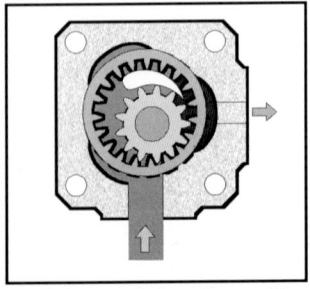

Figure 2. Internal Gear Pump (Eaton)

Internal gear pumps, as shown in Figure 2, have an internal gear driven by the input shaft and an external gear which rotates around its own centre and driven by the internal gear. By means of the separator element, both gears transmit fluid from the pump inlet to the outlet. This pump creates a low noise level that favours it for some applications although its pressure capability is about the same as that of the external gear pump.

3.1.3 Vane pumps/motors

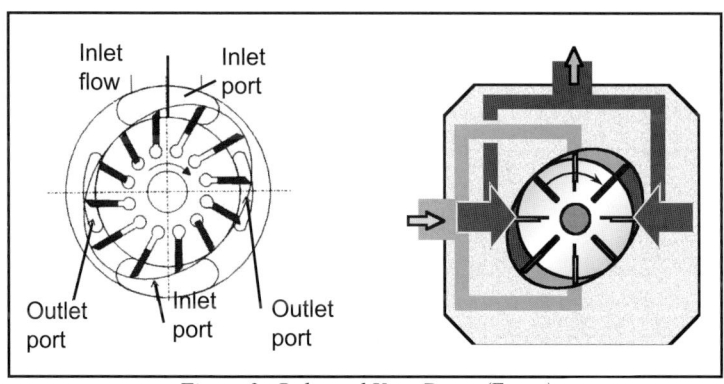

Figure 3. Balanced Vane Pump (Eaton)

The vane pump/motor consists of a rotor, carrying a number of sliding vanes, rotating in a circular housing. With the rotor being eccentric to the casing, oil is transmitted in the vane spaces across the pump from the suction to the discharge port.

The vanes are acted on by centrifugal force when the unit is rotating, but in order to reduce leakage at the tips it is common practice to pressure load them (by supplying discharge pressure to the base of the vane slots) and sometimes to spring load them against the track. As with the gear unit, control of the clearances at the sides of the rotor assembly is most important.

The balanced design in Figure 3 eliminates pressure loading on the bearings and uses an 'elliptical' vane track with the vanes moving in and out twice each revolution. There are diametrically opposed suction ports and discharge ports as shown in Figure 3 and these are connected together in the cast body. This pump is only available as fixed displacement.

Vane pumps are inherently more complex than gear pumps, they contain a greater number of components and are, therefore, more expensive. However, vane pumps operate at much lower noise levels than gear pumps and their cost can be offset against their good serviceability, which is not available with gear type pumps.

4. Variable displacement units

4.1 Vane pumps

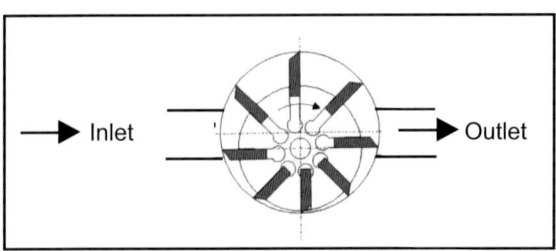

Figure 4. Variable Displacement Vane Pump

Variable displacement vane pumps are available as shown in Figure 4 where the centre of the rotating vane block can be moved in relation to the centre of the housing. Unlike the balanced vane unit of Figure 3, these are single acting and, as a consequence, there is an unbalanced pressure force on the rotor so that the bearing size has to be increased in order to obtain adequate life.

4.2 Piston pumps/motors

Piston units operate at higher efficiencies than gear and vane units and are used for high-pressure applications with hydraulic oil or fire resistant fluids. Several types of piston pump are available that use different design approaches and these include those having axial and radial piston arrangements.

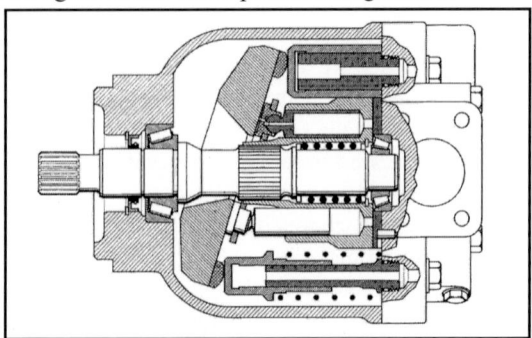

Figure 5. Axial Piston Variable Displacement Pump/Motor (Eaton)

Hydrostatic Pumps and Motors

The majority of piston pumps/motors are of the axial variety, in which several cylinders are grouped in a block around a main axis with their axes parallel as shown in Figure 5 which has variable displacement capability. The pressure force from the pistons is transferred to the angled swash plate lubricated slippers that are mounted onto the pistons with a ball coupling. Rotation of the cylinder block causes the pistons to oscillate in their cylinders by the action of the swash plate, which provides the conversion between the piston pressure force and shaft torque.

The piston cylinders are alternately connected to the high and low-pressure connections by a plate valve between the cylinder block and the port connection housing. Varying the swash plate angle allows the displacement to be changed over the full range from zero to maximum. The swash plate angular position can be arranged to vary either side of the zero displacement position so that flow reversal is obtained. This is referred to as over centre control.

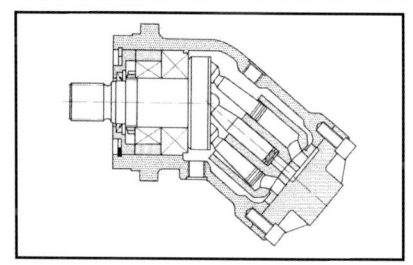

Figure 6. Bent Axis Type Axial Piston Motor (Rexroth)

Figure 6 shows a fixed displacement bent axis type of axial piston unit whereby the ball ended pistons are located in the output shaft. During rotation of the shaft there will be a rotating sliding action in the ball joint, and possibly, between the piston and the cylinder. Each cylinder is connected successively to the high and low-pressure ports by a similar valve to that used in the swash plate units.

In variable displacement units a mechanism is used to vary the tilt angle of the cylinder block from zero to maximum which, if required, can provide over centre operation to give reverse flow.

There are two types of radial piston pump; those in which the cylinder block rotates about a stationary pintle valve, and those with a stationary cylinder block in which the pistons are operated by a rotating eccentric or cam.

Many standard piston units of recent design operate at pressures of up to 450 bar. A wide range of types are available up to powers of 100 kW, although a number of manufacturers provide units having powers up to 300 kW with some available at powers of 1000 kW. Peak overall efficiencies in excess of 90% are usually obtained. The price of piston units varies from manufacturer to manufacturer but may be as much as ten times the price of a gear pump of similar power.

Some pumps cannot draw the inlet fluid directly from the reservoir and may require boosting from a separate pump, often of the external gear type, that can accept low inlet pressures. However, for open loop circuits, variants are available that do not require separate boosting of the inlet which can be connected directly to the reservoir. These aspects also apply to motors that operate as pumps in hydrostatic systems when regeneration occurs (e.g. winch and vehicle drive applications).

Variable displacement pumps provide a range of control methods, which include pressure compensation, load sensing and torque, or power, limiting devices. Pump displacement controls incorporating electro-hydraulic valves are also available.

In addition to their use to control outlet flow, variable displacement pumps provide considerable increase in overall system efficiency with the additional benefits of reduced heat generation and operating cost. This reduction in operating cost can show an overall reduction in the total lifetime cost of the machine.

5. Equations for pumps and motors

5.1 Flow and speed relationship

For the ideal machine with no leakage, the displacement of the machine and its speed of rotation determine the flow rate Q.

Thus:
$$Q = D\omega \tag{1}$$

where D is volumetric displacement [m^3 rad^{-1}]
ω is the rotational speed [rad sec^{-1}]

For pumps that are driven by electric motors the speed is often constant. However for motors, the speed depends on the level of the supplied flow:
Thus:
$$\omega = \frac{Q}{D} \tag{2}$$

5.1.1 Volumetric efficiency

The internal flow leakage in pumps and motors affects the relationship between flow and speed and is taken into account by the use of the volumetric efficiency (η_v).
Thus for pumps equation 1 becomes

$$Q = \eta_v \omega D \qquad (3)$$

And for motors equation 2 becomes

$$\omega = \eta_v \frac{Q}{D} \qquad (4)$$

The volumetric efficiency varies with the fluid viscosity, pressure and rotating speed as discussed in more detail in chapter 8. Manufacturers will usually give values for the volumetric efficiency for operation at specified conditions.

5.2 Torque and pressure relationship

For the ideal machine, the mechanical power is entirely converted to fluid power,

$$\text{Power} = T\omega = PQ \qquad (5)$$

Where T is the torque [Nm]
P is the differential pressure [N m^{-2}]

From equation 5 we get:

$$T = \frac{QP}{\omega} \quad \text{which from equation 2 gives} \quad T = PD \qquad (6)$$

Thus the ideal torque is proportional to the pressure for a given displacement. In a pump this is the input torque required from the prime mover and for a motor, it is the output torque available from the motor shaft.

5.2.1 Mechanical efficiency

The presence of friction between the moving parts creates mechanical losses that are represented by the mechanical efficiency (η_m). Thus:

For pumps the required input torque is given by: $T = \dfrac{PD}{\eta_m}$ (7)

And for motors the output torque is given by: $T = \eta_m PD$ (8)

The mechanical efficiency, as for the volumetric efficiency, will vary with the fluid viscosity, pressure and rotating speed as discussed in more detail in chapter 8. The power input, H, to a pump is:

$$H = \frac{PQ}{\eta_m \eta_v}$$ (9)

The power output from a motor is:

$$H = \eta_m \eta_v PQ$$ (10)

The total efficiency of both units is therefore:

$$\eta_T = \eta_m \eta_v$$

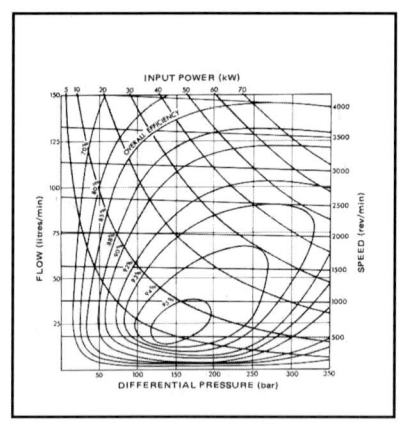

Figure 7. Pump Performance Characteristics

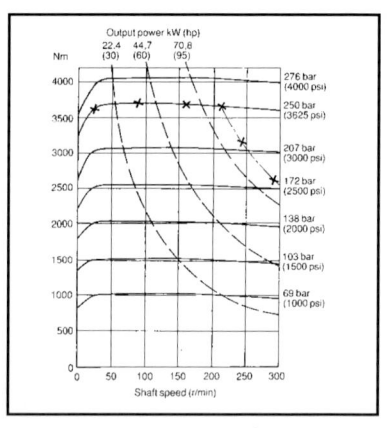

Figure 8. Motor Performance Characteristics

Figures 7 and 8 show how the measured performance of pumps and motors are presented for use with a particular fluid at a particular viscosity. For the pump it can be seen that the flow output reduces with the output pressure at constant speed because of the effect of the increasing leakage flow loss.

For the motor, the output torque varies with increasing speed at constant pressure as a result of the variation in the mechanical efficiency. The theoretical analysis given in chapter 8 shows how the efficiencies are related to the system parameters which enables the performance for operation at other conditions to be predicted.

5.3 Pump selection parameters

The process involved in the selection of a suitable pump for a given application depends on many parameters, some of which were summarised in section 2. As a consequence a generalisation is not possible but some major features can be identified as shown in Table 1.

Gear pump (Fixed displacement)	External type	Internal type	Other features
	•Low cost •Low contaminant sensitivity •Compact, low weight •Good suction performance •250cm^3/rev, 250bar	•Low noise •Low contaminant sensitivity •250cm^3/rev, 250bar	In-line assembly for multi-pump units
Vane	Fixed displacement	Variable displacement	
	•Low noise •Good serviceability •200cm^3/rev, 280bar	•Low noise •Low cost •Good serviceability •Displacement controls •100cm^3/rev, 160bar	In-line assembly for multi-pump units
Piston	Fixed and variable displacement		
	•High efficiency •Good serviceability •Wide range of displacement controls •Up to 1000cm^3/rev, 350/400bar		• Integral boost pump and multi-pump assemblies (not bent axis) •Can use most types in hydrostatic transmissions

Table 1 Comparison of Pump Types

In most cases manufacturer preference and the experience of the machine designer usually dictate the type of pump that is used in applications but in some circumstances it may be necessary to evaluate different types of pump in new applications or where significant changes are required.

6. Low speed motors

As described in the *Introduction*, in principle, most pumps can be operated as motors. However, pumps are low torque high-speed devices and generally require the use of reduction gearboxes in order to provide an output drive at increased torque and reduced speed. The availability of low cost multi-stage gearboxes enables a wide range of ratios to be offered to suit different applications.

In many applications the drive operating speed (e.g. winches, vehicles) is variable in a range from zero to a few hundred revolutions per minute. For this type of application specialist low speed motors are available that have higher operating efficiencies at low speeds. In many cases a low speed motor can be selected that avoids the necessity of employing a gearbox and has sufficient bearing capacity to support side loads on the output shaft.

The range of available low speed motors encompasses a number of different design concepts that include radial piston eccentric and cam, axial piston and those using the Gerotor principle. The motor displacement for a given application is dependent on the required torque and operating pressure. However the type of motor to be used is a function of a number of variables that include; maximum speed, torque and pressure, shaft side load, duty cycle, fixed or variable displacement, weight and cost.

The selection criteria for hydrostatic motors are different from those that apply to pumps because the required output parameter is torque that needs to be available over a wide speed range. This, in most cases extends from zero and may require significant operating periods at speeds below 10 rev/min. Motors are often also required to operate in both directions.

6.1 Types of low speed motors

6.1.1 Radial piston motors

Figures 9 and 10 show two different types of radial piston motor, that of Figure 9 using an eccentric that gives one piston stroke/revolution whilst the cam unit of Figure 10 creates several piston strokes/revolution.

Hydrostatic Pumps and Motors   21

The eccentric motor shown is typical of several designs that are available for converting the piston force into output torque at the drive shaft. In the motor shown, the valve successively connects each cylinder to the supply and return for successive half revolutions of the shaft.

These motors generally operate at a maximum continuous pressure of 250/300bar with displacements up to 10,000cm^3/rev and above.

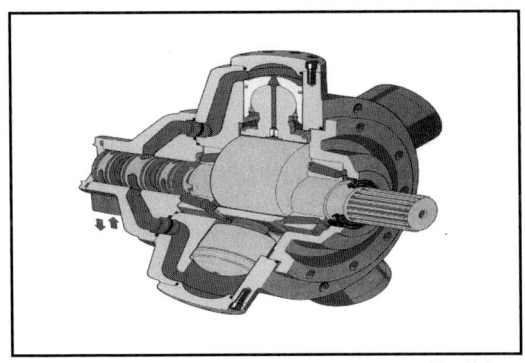

Figure 9. Radial Piston Eccentric Motor (Staffa)

The motor displacement can be altered between two levels by pressurising the pistons that control the position of the eccentric. In some motors the displacement can be controlled continuously in a closed loop control on inlet pressure or motor flow.

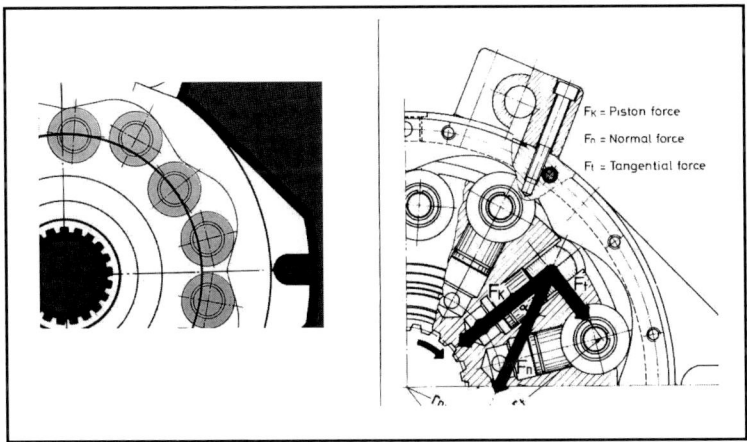

Figure 10. Radial Piston Cam Type Motor (Hydrex)

The particular type of radial piston cam motor shown in Figure 10 operates by transferring the pressure force on the piston, which is directed radially outwards, onto the cam by the use of rolling element bearings attached to the piston. Here the distributor valve, not shown in the figure, connects each piston to the high-pressure port when the piston is moving outwards thus creating a torque on the cylindrical cam. When the piston is moving inwards the valve connects the cylinder to the low-pressure port thus allowing fluid to be passed back to the return line.

Radial piston cam motors generally operate at a continuous pressure of 250/300bar with displacements up to, and beyond, 50,000cm^3/rev and can be operated with two selectable displacements by short circuiting some of the cylinders to low pressure. Cam motors can be arranged to provide low levels of torque ripple and they have a high ratio of output torque to the motor mass (see Figure 13).

In the Gerotor, or Orbit type motor, as shown in Figure 11, the rotor centre rotates in an orbit as the rotor rolls in contact with the outer rim. This action causes the volumes trapped between the several contacts to vary with rotation of the shaft.

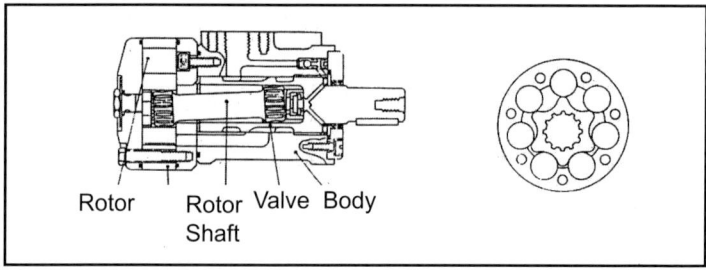

Figure 11. Gerotor, or Orbit Type Low Speed High Torque Motor (Danfoss)

The rotating valve connects increasing volumes to the high pressure port and reducing volumes to the low pressure port this action creating a torque on the output shaft and a continuous rotation in accordance with the supply flow.

Orbit motors have a high ratio of torque to motor mass but operate at lower pressures than the radial piston motors and are usually employed for medium pressure, low power applications. Typically motor displacements are up to 800 cm^3/rev at pressures up to 175/200bar.

7. Some general considerations

To avoid motor malfunction the system must be capable of working with the range of loads and speeds, both steady and transient, that the application demands. Some general points to be considered are as follows:

Hydrostatic Pumps and Motors   23

- Displacement control - The operation of motor displacement variation has to be carefully considered for a given application as the pressure and speed levels are determined by the ratios of $\frac{T}{D}$ and $\frac{Q}{D}$ so that for given values of torque and flow, reducing the displacement increases both the pressure and the speed.
- Motor creep - A motor under load (e.g. winch) with closed ports will rotate slowly backwards (creep) because of internal leakage.
- Torque ripple - This refers to variations in the shaft torque whilst rotating. It is only of significance at speeds where the ripple frequency is in the range of zero to slightly higher than the natural frequency of the hydraulic system.
- Backpressure - Some motors cannot operate for significant periods of time with high backpressure (i.e. working as a pump in meter-out mode).
- Boost pressure – Some motors require a certain level of boost pressure in order to keep the mechanical parts together when operating at high speed. This is particularly important, for example, during overrunning conditions (e.g. winch systems) when cavitation can occur.
- Motor drain - The external drain from the motor should allow completely free flow in order to avoid excessive pressure levels occurring in the crankcase or sump.

8. Comparison of Motor Performance Characteristics

Type	Max Displacement (cm³/rev)	Max Pressure (bar)	Max Torque (Nm)
External gear	250	250	1000
Vane	350	250	1400
Orbit	800	175	2200
Radial piston eccentric	65000	300	310000
Radial piston cam	100000	300	480000

Table 2 Motor Data

As can be seen from the foregoing discussion, there is a large variety of motor types that can be used for any given application that includes high speed, low torque (HSLT) and low speed, high torque (LSHT) units. Table 2 gives a summary of the nominal maximum displacement and torque for these types of motors.

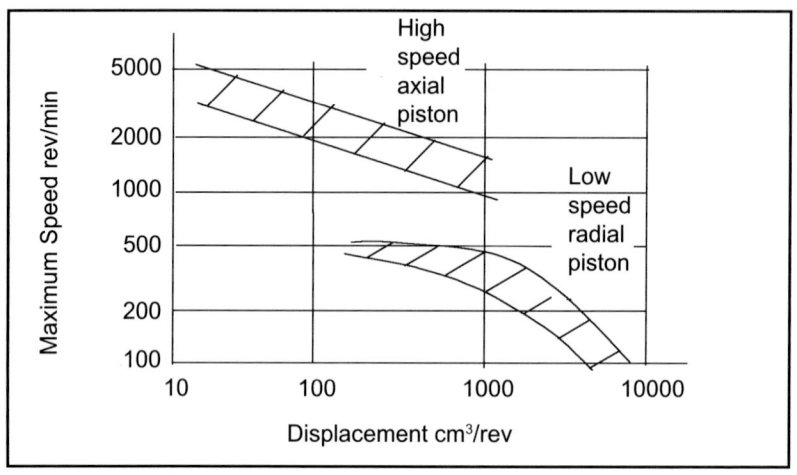

Figure 12. Typical Maximum Speeds for Motors of Various Types

HSLT motors generally have a higher speed capability than LSHT motors as seen in Figure 12 (e.g. axial piston swash plate/bent axis cf. radial piston motors). In general, a gear reducer ratio of between 7 and 20:1 will provide HSLT motors with torque levels that are similar to those obtained from LSHT motors. This can be obtained from a single stage epicyclic gearbox (max. ratio of 7:1) and a single ratio spur gear set or a two stage epicyclic gear unit.

The mass of such a combined drive unit can be generally similar to those of LSHT motors. The space envelope of these units is longer and smaller in diam-

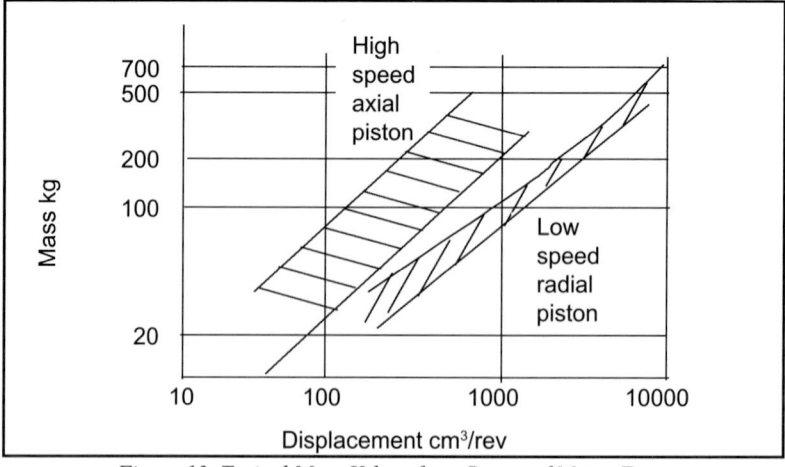

Figure 13. Typical Mass Values for a Range of Motor Types

eter than that of radial motors and may be a disadvantage in some applications such as vehicle drives and some industrial systems. However, for winch drives the gearbox can be installed inside the winch drum which provides an optimal solution to system design in some applications.

When compared directly on the basis of displacement LSHT motors generally are of lower mass thus giving higher values of specific torque. Typically this parameter ranges from 5 to 20 Nm/kg for HSLT motors and 20 to 70 Nm/kg for LSHT motors. However, LSHT motors generally have lower values of specific power with values up to 3kW/kg as compared to 8kW/kg for HSLT motors.

There is a wide range of motor drive units from which to select for a given application but the choice may require the consideration of cost (capital and operating), machine design aspects such as its control, servicing and maintenance and a user preference which is usually based on previous experience.

Reference

P J Chapple, A performance comparison of hydrostatic piston motors - factors affecting their application and use - 7th BHRA International Fluid Power Symposium 16/18 September 1986.

CHAPTER THREE

HYDRAULIC ACTUATORS

3. HYDRAULIC ACTUATORS

Summary

As discussed in chapter 2 the flow output from a pump can be used to drive a motor where the rotary speed of the motor is determined by its displacement. Hydrostatic motors are a class of actuator in that they convert flow and pressure into velocity and torque on a continuous basis but they can also be used in some applications where only limited movement is required.

However, rotary actuators are available that have a limited rotation angle and offer a reduction in cost because of their relative mechanical simplicity. These can also avoid the need for a holding brake because of their low leakage. Hydraulic cylinders are normally used for providing linear motion, which can have zero leakage so that, with blocked ports, stationary loads will be held indefinitely. The use of a rack and pinion gear drive allows the linear movement to be converted to rotary motion whilst retaining the stalled characteristics of the hydraulic cylinder.

Linear actuators, or cylinders, are extensively used in all of the major engineering fields and the system designer needs to be aware of the different types of construction that are available, the various mounting methods and the influence that these have on their load carrying characteristics.

1. Introduction

This chapter is concerned with:

- The construction, mounting methods and cushioning of linear actuators.
- The construction of the major types of rotary actuators

2. Linear actuators

Linear actuators convert flow into linear movement and pressure into force by employing a piston that slides inside a cylinder. The construction of a typical double-acting actuator can be seen in Figure 1 that shows the use of appropriate sliding seals for the piston and rod components. The double-acting feature to provide operation in both directions is not always required as in some applications (e.g., fork lift truck systems) the force from the load is used for retraction.

Actuators are also available that have a rod at both ends so that the piston areas are equal in both directions of movement.

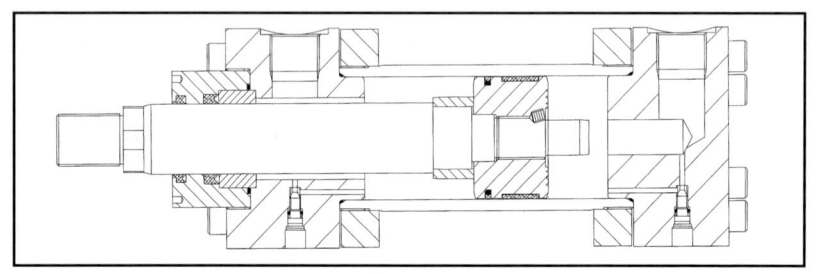

Figure 1. Typical double-acting actuator

3. Principle features

3.1 End covers

The basic methods of attaching the end covers, or caps, to the cylinder of hydraulic actuators include:

- Screwed to the cylinder. This is the method shown in Figure 1.
- Tie rods as shown in Figure 2.
- Welded

3.2 Mounting methods

A wide variety of actuator and rod end mounting methods are available to suit the requirements of different applications. Figure 2 shows a flange mounting that is part of the front end cap for locating the actuator rigidly to the machine frame.

Hydraulic Actuators

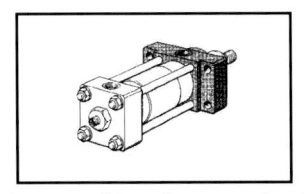

Figure 2. Actuator of tie rod construction (Mecman)

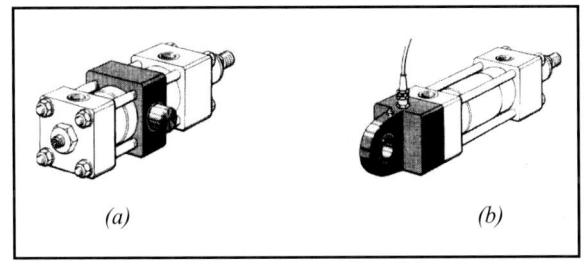

(a) *(b)*

Figure 3. Actuator mountings (Mecman)

The trunnion mounting shown in Figure 3a provides a pivoting action in one plane whereas the mounting shown in Figure 3b uses a spherical bearing which allows pivotal movements in any plane. As is discussed the mounting style used has a significant influence on the actuator strength and the various alternative methods are shown in Figure 4.

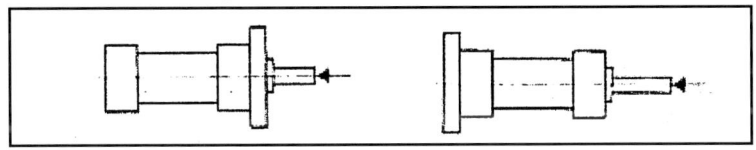

a) Actuator flange mounted to front or rear end cover with the rod end unguided

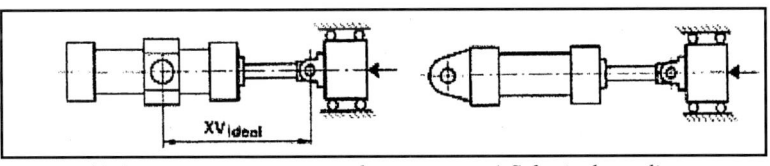

b) Actuator trunnion mounted with the rod end guided   *c) Spherical coupling with the rod end guided*

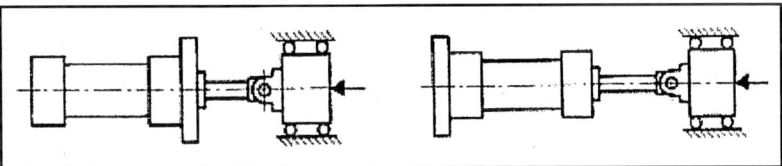

d) Actuator flange mounted with the rod end guided
Figure 4. Actuator mounting styles (Mecman)

For working against pushing load forces the actuator acts as a strut for which the Euler failure criteria are applied according to the method of mounting the strut. The Euler buckling load, F_E, is given by:

$$F_E = \frac{S_F \pi^2 E I}{L^2}$$

where:

E = Youngs Modulus

I = Second Moment of Area = $\frac{\pi d^4}{64}$

d = Rod Diameter

L = Rod Length

S_F = Actuator Strength Factor

Thus it is seen that for a given strength factor the buckling load varies proportionally with the fourth power of diameter and inversely with the length squared.

The values for the strength factor, S_F, that apply to the mounting styles in Figure 4 are:
i) Fixed actuator mounting with unconstrained rod end (as a) $S_F = 0.25$
ii) Actuator and rod attached by free pivots but with constrained rod end (as c) $S_F = 1$
iii) Fixed actuator mounting with constrained rod end (as d) $S_F = 2$

Actuator manufacturers usually give the maximum capability of actuators with the mounting style in terms of maximum extension at a given actuator piston pressure.

A typical example is given in Table 1 for an actuator of 50mm diameter at 100 bar pressure.

d (mm)	Maximum Piston Extension (mm)			
	Case a)	Case b)	Case c)	Case d)
28	390	610	500	1260
36	690	1120	730	1690

Table 1 Maximum Actuator Extension (refers to Figure 4)

3.3 Seals

Static sealing is normally achieved by O-seals trapped in grooves of the appropriate size, although gaskets are sometimes used. There should be no leakage from these devices.

Dynamic sealing, particularly of reciprocating pistons or rods, involves leakage and friction. Proprietary synthetic seals or packing can give almost perfect sealing since they are pressure loaded. Types commonly used are cup seals, U-rings, V-rings or Chevron packing, and composite rings using different polymers which can also provide load bearing capability.

The problem with seals is friction, particularly static friction (especially after an idle period), and friction at high pressures when the seal is deformed. Today, modern seals can have low friction properties when combined with other seal materials. The material that is recommended for the seals in any given application would depend on the type of fluid used, its temperature and the maximum actuator velocity.

Rod seals must not only prevent the leakage of fluid but also prevent the ingress of dirt. Wiper rings are normally incorporated into the end cover and bellows units, or gaiters, may be fitted to piston rods that are operating in unfavourable environments.

In practice, many piston rods are subject to lateral as well as axial loads. These loads must not be permitted to cause piston misalignment inside the cylinder. The seal housing fitted to the end cover may be provided with a bearing bush to carry side loads on the rod.

3.4 Position transducers

Many manufacturers include position transducers in the actuator assembly. This avoids having to fit an external transducer that, in many applications, can be exposed to the possibility of damage. The methods used for position indication vary but usually incorporate transducers involving no physical contact, (e.g. inductive and ultrasonic systems) giving digital or analogue output.

4. Actuator selection

4.1 Actuator force

For the maximum load force (stall force), the system pressure and actuator size may be determined. Factors to consider when choosing system pressure are the mounting style (as discussed in 3.2), duty cycle, utilisation, performance, reliability and cost.

For an ideal actuator,

$$Force = \frac{Pressure}{Area}$$

However in practice various losses must be allowed for:

i) There will normally be pressure on both sides of the piston.

$$Net\ force = P_H A_H - P_L A_L$$

where suffices H and L refer to the high and low pressure sides.

ii) The force available at the load is reduced by friction. Frictional forces are difficult to predict, but may well be of the order of 20% of the load under operating conditions and higher for starting conditions.

iii) An allowance should be made for the efficiency of any mechanical linkages or gears connected to the output.

iv) In valve controlled systems (e.g., meter-in), the inlet pressure will reduce with increasing actuator velocity. Thus, for systems in which the force is varying, the velocity may vary as a consequence.

In applications where the load is unguided, transverse loads may be a acting. A stop tube, or spacer, is sometimes fitted to reduce the stroke in such instances but in any case if such loads are to be expected on the actuator the application should be discussed with the actuator manufacturer.

4.2 Cushioning

To retard inertial loads and increase the fatigue life of actuators, some form of internal cushioning is often used. An example of actuator cushioning can be seen in Figure 5.

When the actuator outlet flow is directed through the restrictor, the pressure drop generated will create a backpressure on the actuator, thus causing it to be retarded. The restrictor must be sized such that the maximum pressure, which occurs when the plunger first blocks the normal outlet port, does not exceed the safe value for the actuator.

Hydraulic Actuators

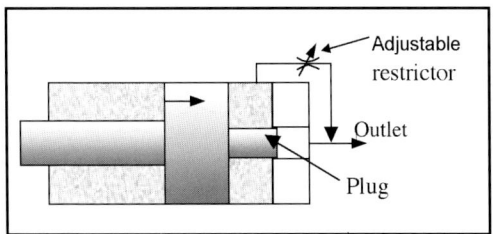

Figure 5. Actuator cushioning

For simple inertial loads, with no other forces acting, the actuator velocity decays exponentially, as does the actuator outlet pressure. This can be shown by simple analysis assuming an incompressible fluid and neglecting friction. Thus from Newton's Law we have:

Inertial force

$$m \frac{d^2 X}{dt^2} = m \frac{dU}{dt} = -P_C A_C \quad (1)$$

where X is the movement of the actuator from the commencement of cushioning.

Flow

Restrictor
$$Q_R = C_D A_R \sqrt{\frac{2}{\rho} P_C}$$

$$\therefore P_C = \frac{Q_R^2}{C_D^2 A_R^2} \frac{\rho}{2} \quad (2)$$

Actuator:
$$Q_C = Q_R = A_C U \quad (3)$$

Now: $\frac{dU}{dt} = \frac{dU}{dX} \frac{dX}{dt} = U \frac{dU}{dX}$ so we get from equations (1), (2) and (3):

$$P_C = \frac{U^2 A_C^2}{C_D^2 A_R^2} \frac{\rho}{2} = -\frac{mU}{A_C} \frac{dU}{dX} \quad (4)$$

And:

$$\int_{U_m}^{U} \frac{dU}{U} = -\frac{\rho A_C^3}{2C_D^2 A_R^2 m} \int_0^X dX = -\frac{C}{m} \int_0^X dX \qquad (5)$$

Solution

The solution of equation (5) gives:

$$U = U_m \exp\left(-\frac{CX}{m}\right)$$

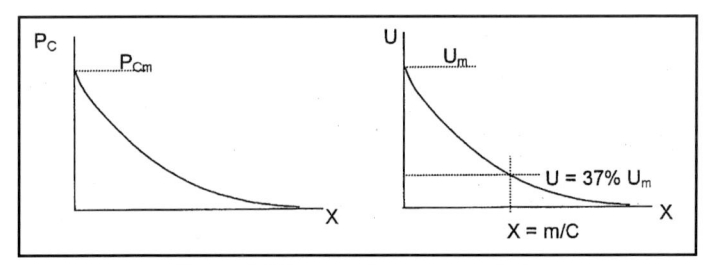

Figure 6. Velocity and pressure variation

The velocity and pressure variations with the distance, X, which the actuator has moved after cushioning has commenced are shown in Figure 6. At the start of the cushioning the pressure rises to a maximum value, P_{Cm}, when the flow is a maximum. For a given mass and initial velocity, the maximum cushion pressure is determined by the size of the adjustable restrictor. This also determines the distance that is required for the actuator velocity to reduce to an acceptable value.

It is normal that P_C max should not exceed 350bar which is a normal fatigue pressure rating for 10^6 actuator cycles. The change in the pressure, and velocity, will be slightly modified by the effect of the fluid compressibility but in most systems this effect will be small and the cushion performance can be calculated using the equations.

Some cushioning systems employ a long tapered plunger that maintains a higher mean pressure throughout the cushioning stroke and, consequently reduces the cushioning distance. In others the plunger has stepped diameters to give almost the same effect. The performance of these cushion methods is usually given in manufacturer's literature in terms of the energy that is to be absorbed and the pressure level on the supply side of the actuator.

The shortest cushion length would be obtained from one that creates a constant pressure at the maximum permissible value. The performance of some cushion systems can be found in the papers by Chapple[1] and Lie, Chapple and Tilley[2]. A comparison with a tapered cushion shows that the cushion length can be reduced by around 30%.

5. Rotary actuators

Rotary actuators are designed specifically to provide a limited angle of rotation. These are distinct from hydraulic motors and as a consequence are of simple design as no timing valve is necessary.

5.1 Actuator types and capacity range

Three common types rotary actuator: rack and pinion, vane and helical screw are shown in Figures 6, 7 and 8. A summary of their performance is given in Table 2.

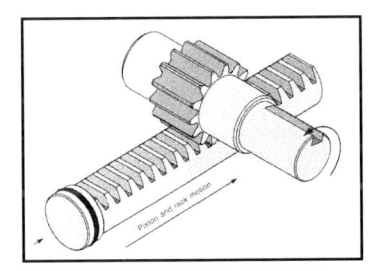

Figure 7. Typical rack and Pinion Rotary Actuator

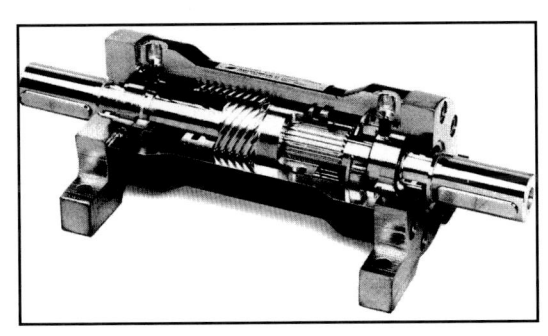

Figure 8. Screw Type Rotary Actuator (Danfoss)

Type	Angle range	Torque Nm
Rack and pinion	> 360°	42000
Vane	< 280°	22000
Helical	< 420°	26000

Table 2. Summary of rotary actuator performance

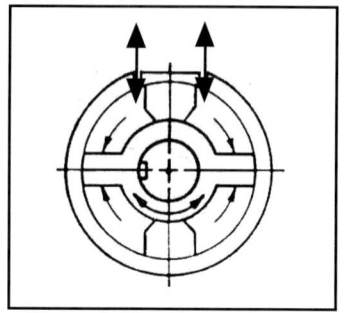

Figure 9. Vane Rotary Actuator

5.2 Applications

Rotary actuators are used for the following applications:
- Steering systems
- Manipulator drive
- Gate valves
- Tunnelling machine
- Boom slew of backhoe
- Container handling

Most of the actuators will carry side loads and can usually be supplied with position indication, cushioning valves, mechanical stops and a variety of shaft attachment features.

References

1. **P J Chapple**, Using simulation techniques in the design of actuator cushioning, Drives and Controls Conference, Kamtech Publishing, Telford, UK, March 1999.

2. **T Lie, P J Chapple and D G Tilley**, Actuator cushioning performance, simulation and test results, PTMC Conference, University of Bath, Bath, UK, September 2000.

CHAPTER FOUR

PRESSURE CONTROL VALVES

4. PRESSURE CONTROL VALVES

Summary

Generally the pressure level in hydraulic systems will vary so as to provide the required torque or force from an actuator in order to drive an external load from the particular application. During start and stopping situations and when the load is varying transiently the pressure may exceed the maximum safe value for the system. In many systems several actuators will be driven by a single pump and, in addition to limiting the supply pressure it may be necessary to reduce the pressure level supplied to individual services. There are many different types of valves on the market and this chapter describes some of these and the operating principles involved.

1. Introduction

Major types of valves that are used to control the pressure level in hydraulic systems include:

- Relief valves for limiting the maximum system pressure
- Reducing valves for limiting the pressure in parts of a circuit at a lower level than in the supply system
- Load control valves to control the motion of an actuator or motor under the action of overrunning, or negative, forces.

The type that is employed in a given application depends on the particular requirements and system specification.

2. Relief valves

Relief valves are the most commonly used pressure control valve as they are required in all systems to prevent the generation of excessive pressures. In many systems they are used in combination with a pump to provide a source of flow at constant pressure.

2.1 Single stage relief valve

Single stage relief valves, as with all valves, can use either a piston, or a spool, that is opened by a pressure force against a preloaded spring as shown in Figure 1 together with the ISO symbol that is used to represent it in a circuit diagram. On opening, the valve allows some of the supply flow to be passed back to the tank thus limiting the maximum pressure in the supply. The valve needs to be sized such that all of the supply flow can be returned to tank at a supply pressure that does not exceed the maximum desired level.

Single stage relief valves are the simplest and lowest cost valve. Considering the diagram of a poppet valve in Figure 2, the valve will start to open at its 'cracking pressure' when the force on its face due to the inlet pressure is equal to the spring preload.

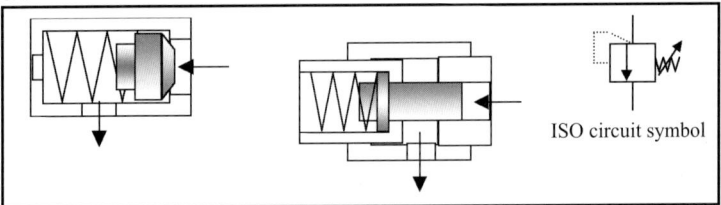

Figure 1. Poppet and Piston Type Valves

As the pressure increases above this value the valve opens progressively thus allowing flow to pass through the valve. This feature is referred to as the 'pressure over-ride'.

The rate of the spring will determine the relationship between the valve displacement and the inlet pressure. In order to minimise the free length of the spring requires the spring to have a high stiffness. However, the higher the stiffness the greater will be the pressure over-ride. There will be some hysteresis due to friction between the components that will result in the pressure being slightly higher when the valve is being opened than when it is being closed.

As is discussed in Chapter 8 there is an additional force that arises from the increase in the momentum of the fluid as it passes through the valve opening

Pressure Control Valves 43

which acts in the direction to close the valve. This force acts as a spring the rate of which is added to that of the mechanical spring and its effect is to increase the pressure override.

2.2 Two Stage Relief Valves

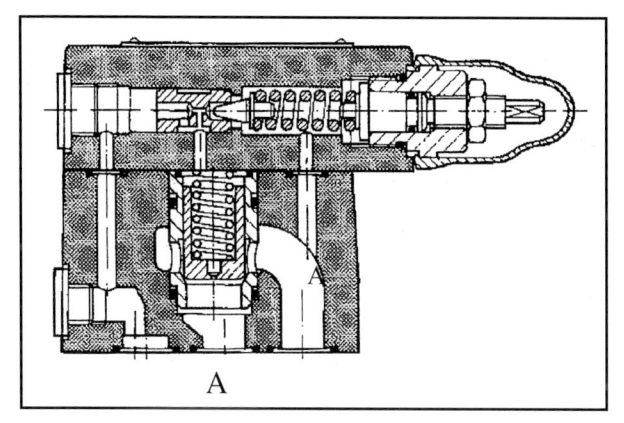

Figure 2. Two Stage Poppet Type Relief Valve (Eaton)

The two stage valve shown in Figure 2 uses a spring loaded pilot poppet (pilot relief valve) to sense the pressure level in the supply at A. When this pressure causes the pilot relief valve to open the flow through the balancing orifice creates a pressure drop across the main valve poppet that has a spring preload in the region of 2 bar. This causes flow from the supply to be returned to tank at a controlled level of the supply pressure.

Two stage valves have a much reduced pressure over-ride as compared to single stage valves because the main spring is not required to be preloaded to the controlled pressure level. This allows a reduced spring rate to be used that reduces the pressure over-ride. This can be of advantage where it is required to control the supply pressure within close limits.

The pilot relief valve can be isolated from the main valve. There can be more than one pilot valve which can be set at different pressures so that the connection of any one will operate the relief at the respective pilot set pressure.

Figure 3 shows a typical cartridge type of two-stage relief valve. Cartridge valves are available for installation in individual housings, sandwich mounting (stacking) and in special manifold blocks.

Relief valves are also available with electric control using proportional solenoids to provide the necessary force in place of a mechanical spring. A major

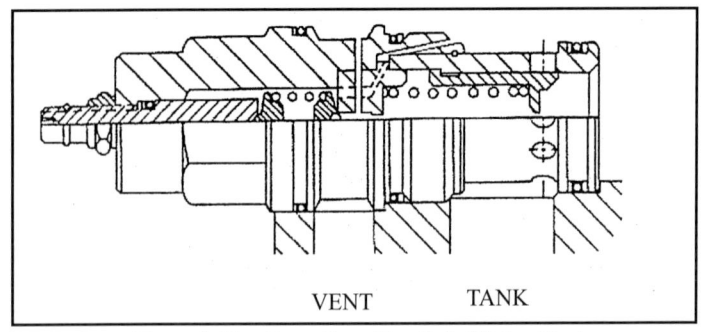

Figure 3. Two Stage Cartridge Relief Valve (Sun)

advantage of this type of valve is the ability to adjust the pressure setting from an appropriate voltage controlled amplifier.

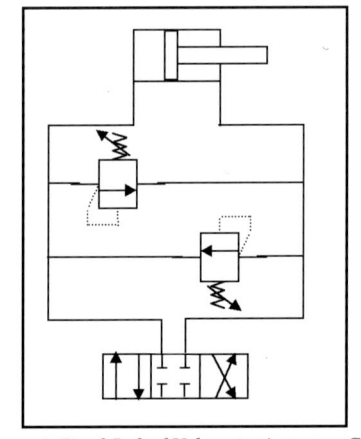

Figure 4. Dual Relief Valves in Actuator Circuit

Dual crossline relief valves, as shown in Figure 4 for limiting the pressure on both sides of a linear actuator, are available in a single casing.

3. Pressure Reducing Valves

A pressure reducing valve is used to provide a sub-circuit with a supply of fluid at a pressure which is less than the pressure in the main circuit. Figure 5 shows a schematic diagram.

Pressure Control Valves 45

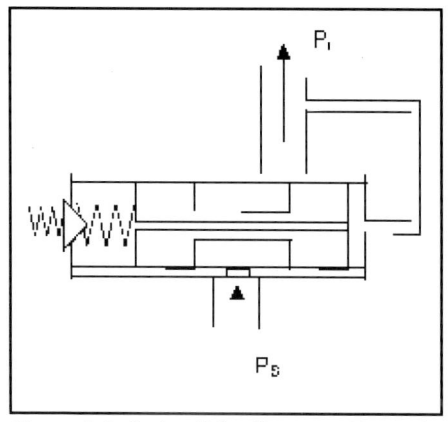

Figure 5. Reducing Valve Operating Principles

The downstream pressure P_r acts on the first-stage, spring loaded, poppet valve and when this valve is open, the resultant flow through the orifice drilled through the spool valve creates a pressure differential between the two ends of the spool. This moves the spool against a spring so that the spool throttles the flow between the supply and service ports. If the downstream pressure rises above the required level, the spool moves to increase the throttling action (and vice versa).

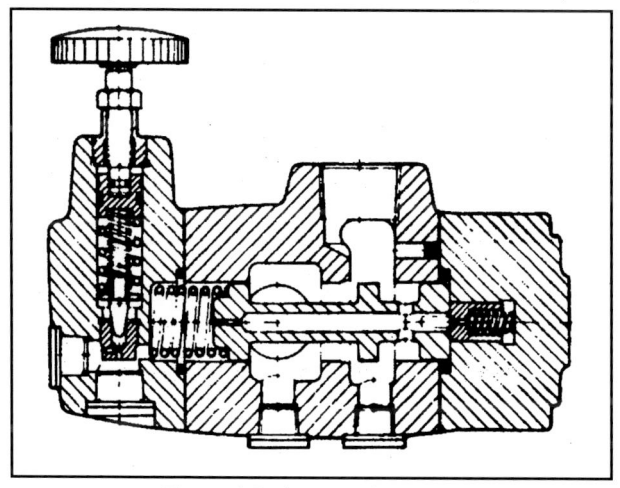

Figure 6. Reducing Valve (Eaton)

The reducing valve in Figure 6 has a screw adjustment to change the set pressure.

4. Counterbalance Valves

For actuators, or motors, under the action of negative forces (e.g. pulling forces during extension) it is necessary to provide a restriction in the flow outlet in order to create a resisting, or back, pressure.

Load control valves of the counterbalance type are shown in Figure 7. That in (a) is a line mounted valve where the valve opens when the force from the pilot

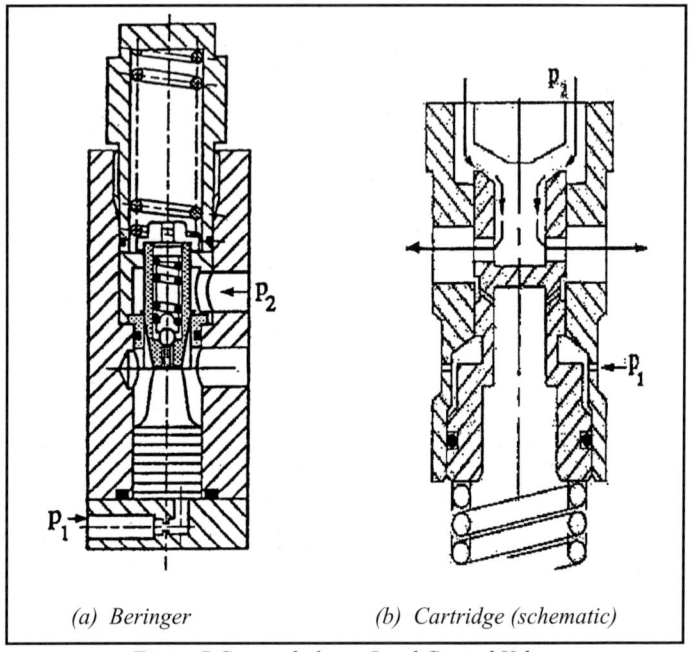

(a) Beringer   *(b) Cartridge (schematic)*

Figure 7 Counterbalance Load Control Valves

pressure, P_1, exceeds that of the spring. The operation of the valve is independent of the inlet pressure P_2 whereas for the cartridge valve in (b) this is not the case as the valve opening is controlled by both of the pressures. The ratio of the areas exposed to the pressures P_1 and P_2 can be selected this usually being in the range 3:1 and 10:1.

A typical operating characteristic of valve type (b) is shown in Figure 8 for the cracking pressure and for two flows. The flow through the valve depends on its opening and the pressure P_2 so as P_2 reduces, the valve opening has to be increased which requires P_1 to have a higher value over that for the valve just cracking open. This effect can be seen from Figure 8.

Pressure Control Valves 47

The maximum value of P_2 which is set by pre-loading the spring with a screw adjustment, is chosen to provide a maximum safe pressure for the actuator and occurs when P_1 is zero. For the actuator the variation of P_2 with P_1 depends on its area ratio which has a minimum value of unity for an equal area cylinder.

For a given force on the actuator the pressure relationship for extension of the actuator is shown in Figure 8 superimposed onto the valve characteristics, the point of intersection providing the operating condition of the system. Standard cylinders normally have a maximum area ratio of around 2 but some applications use actuators having ratios considerably higher than this. Clearly this has an influence on the operating pressures when under the control of a counterbalance valve.

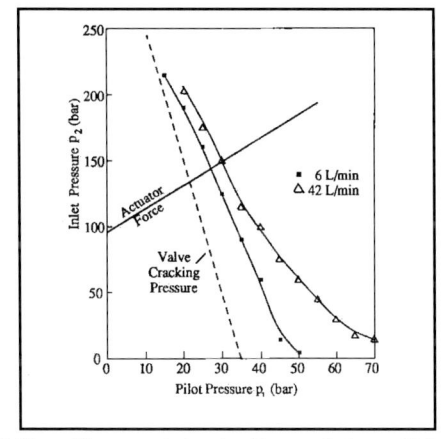

Figure 8 Flow Characteristics for Counterbalance Valve type (b)

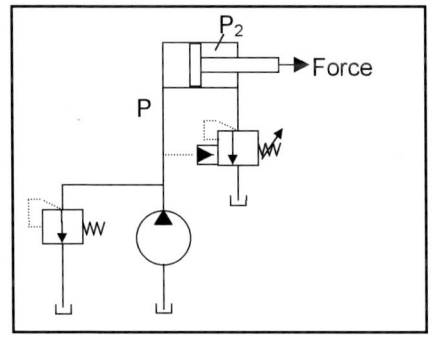

Figure 9 Extending Actuator Controlled by a Counterbalance Valve with a Pulling Load

As seen from Figure 9 the pilot pressure, P_1, is obtained from the pump outlet to the actuator inlet when the actuator is extending. If the actuator flow exceeds that of the pump P_1 will reduce and close the valve. This will cause the actuator outlet pressure, P_2, to increase and thus reduce the actuator speed until the pressures create an opening of the valve that provides equality of the actuator and pump flows and also provides a pressure force that is equal to the load force.

The dynamic performance of counterbalance valve systems, which operate as closed loop systems because of the use of pressure feedback to control the valve position, is extremely complex and can result in oscillatory motion of the load. For large systems it may be advisable to carry out a simulation of the system in order to avoid this problem.

The operating pressures can be estimated for a given valve and actuator using the area ratios for these components and the maximum pressure setting, P_s, so, considering a valve having an area ratio of a the pressures to just crack the valve open are given by:

$$P_1 = \frac{P_S - P_2}{a}$$

For the actuator the value of P_2 when $P_1 = 0$ is $P_2 = P_L = \frac{Force\ F}{Piston\ Area}$. For the force balance of the actuator the pressure P_2 will increase with changes in P_1. Thus, for an actuator area ratio of α:

$$P_2 = P_L + \alpha P_1 \qquad \therefore \qquad P_1 = \frac{(P_2 - P_L)}{\alpha}$$

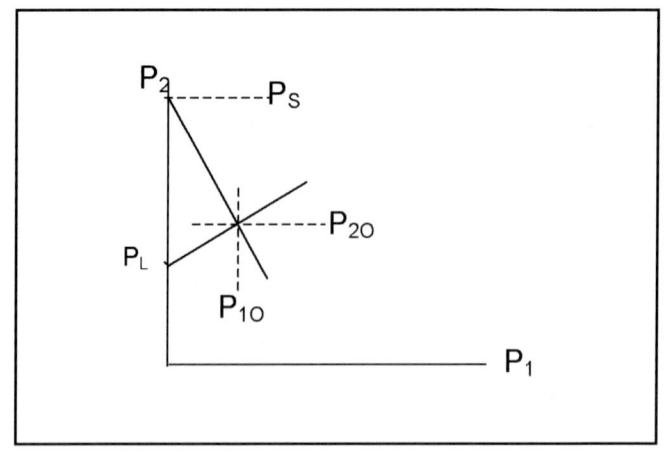

Figure 10. Operating Pressures

Pressure Control Valves

Thus referring to Figure 10, the value of the operating pressure P_{20} can be obtained from equating the two values of P_1 :

$$\frac{P_S - P_{20}}{a} = \frac{(P_{20} - P_L)}{\alpha}$$

Therefore

$$P_{20} = \frac{aP_L + \alpha P_L}{a + \alpha} = \frac{P_L + \alpha/a \, P_S}{1 + \alpha/a}$$

For a very high valve area ratio, $a \to \infty$, $P_{20} \to P_L$.

Thus for $P_S = 200\,bar$, $P_L = 100\,bar$, $\alpha = 2$ and $a = 10$ gives $P_{10} = 8.3\,bar$ and $P_{20} = 117\,bar$.

CHAPTER FIVE

FLOW CONTROL VALVES

5. FLOW CONTROL VALVES

Summary

The flow from hydraulic pumps can be fixed or variable depending on the type of pump being used. For fixed displacement pump systems the simplest form of flow control is obtained by generating a level of pump pressure that causes operation of the relief valve but this method can be extremely inefficient. Alternative valve systems can be used to improve this situation particularly when operating several services from a single pump. The type of circuit used in this case will depend on whether the pump is fixed or variable diaplacement. Some of the available circuit options are described in chapter 7 that use valve types that are described in this chapter.

1. Introduction

The control of flow is broadly divided into major types that include
- Directional control
- Simple restrictor
- Pressure compensated
- Open centre and bypass

The control of flow is a major feature of hydraulic systems and there are a variety of methods available for this which can be used with both fixed and variable displacement pumps. This chapter is concerned with the basic aspects of flow control valves and their characteristics that are of importance in the design and selection of hydraulic circuits.

2. Directional control valve

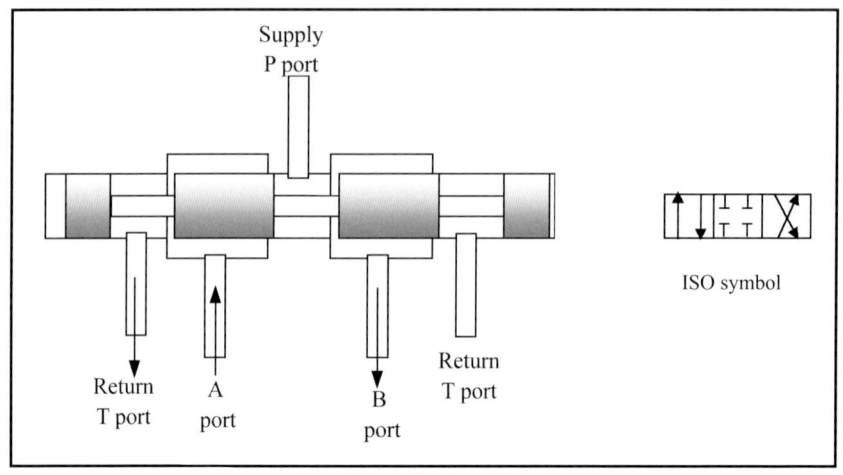

Figure 1 Four Way Directional Control Valve

Spool type valves provide the major method of controlling the direction of flow and are used extensively for many circuit functions that are discussed in the chapter concerned with circuit design. Figure 1 shows the basic features of this type of directional control valve (DCV) which connects one of the two outlet ports (B) to the supply port (P) and the other (A) port to the tank or return line with movement of the spool from the central position.

The valve of Figure 1 has all the ports closed in the centre position as represented by the ISO symbol however, as discussed in chapter 7, other configura-

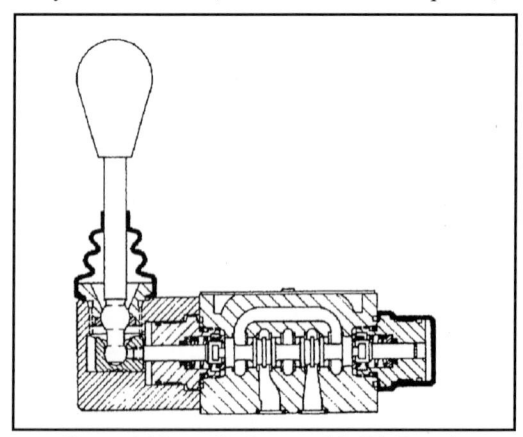

Figure 2 Manually Operated DCV (Eaton)

tions having a range of spool options and port connections are available in order to perform a range of alternative circuit functions. For holding the valve in given positions, they can be spring centred and have detents that engage at particular spool displacements.

The valve can be positioned by a direct manual control, as shown in Figure 2 where the input lever is connected by a spherical coupling to the end of the spool. Other means of positioning the spool are available including hydraulic or pneumatic pilot signals, direct force control from an electric solenoid or indirectly from a solenoid using a hydraulic amplifier (e.g. electrohydraulic servovalve).

Pilot control has a distinct advantage over mechanical operation because the signal can be derived some distance from the valve itself. The electrically operated proportional valve, shown in Figure 3, has high accuracy and resolution and these valves are used extensively in a wide range of applications for both open and closed loop control of actuator position.

The response time of proportional valves is dependent on the type and size of the valve and is quoted in the manufacturer's technical literature. The rate at which these valves open and close can be adjusted in the amplifier in order to minimise the magnitude of pressure shocks in the system. This is particularly applicable to loads that have a large mass or inertia.

Monitoring signals are available so that failures can be detected and the valve put into a system safe position. The maximum flow is limited by the effect of flow forces, discussed in chapter 8, which oppose the force from the solenoid. It is important to observe the manufacturer's recommendation on contamination control, as fluid borne particles are the major cause of failure and unreliable operation of control valves including solenoid burnout in AC valves.

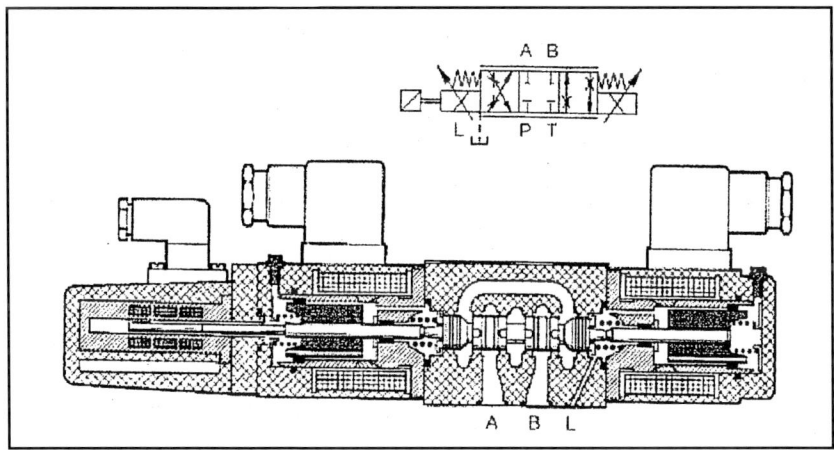

Figure 3 Proportional Control Valve

3. Restrictor valve

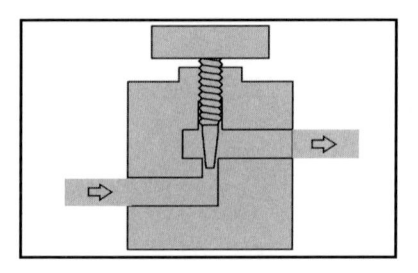

Figure 4. Adjustable Restrictor Valve

Spool valves, particularly the proportional type are used extensively for restrictive control of actuators as discussed in chapter 7 on circuit design. A simple method of flow control can be obtained using a restrictor for meter-in, meter-out and bypass circuits. An adjustable type is shown in Figure 4 where the position of the tapered needle is altered by the screw. Pressure is created at the valve inlet in order to pass the flow through the tapered orifice.

Restrictors create a pressure loss in the flow as described in chapter 8 for orifices and are used with a controlled pressure source. The flow through the valve will vary with the square root of the available pressure drop and the adjustment allows for obtaining the desired flow in an application.

The restrictor valve is normally preset but variable control can be obtained by positioning the spool in directional control valves to give a range of openings from closed to fully open. Proportional valves provide very accurate control of spool position and are often used to vary the restriction in the inlet and outlet flow paths to an actuator as is described in chapter 7.

In situations where the pressure drop is not constant the flow will vary as a consequence. To avoid this problem pressure compensated valves can be used as described in the next section.

4. Pressure compensated valve

Pressure compensated valves use an adjustable restrictor together with an additional valve that opens and closes in order to maintain a constant pressure drop across the restrictor.

Figure 5 shows this type of valve whereby as flow passes from the A to the B ports the pressure drop, $(P_2 - P_3)$ across the manually adjustable rotary restrictor creates a force on the spool against that of the spring. If the force from the pres-

Figure 4. Adjustable Restrictor Valve

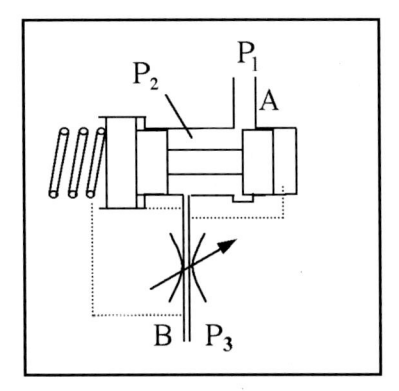

Figure 5. Pressure Compensated Valve

sure drop exceeds the value set by the spring the spool valve closes and further restricts the outlet flow to the B port. With the inlet being supplied from a controlled pressure source, P_1, the supply flow will be reduced in order to maintain this pressure at a constant value. Either a pressure relief valve or a pressure compensated pump could be used to perform this action.

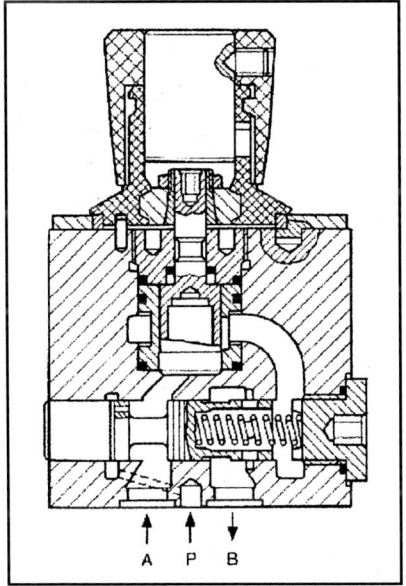

Figure 6. Pressure Compensated Flow Control Valve (Eaton)

Thus the flow is kept constant with changes in the outlet pressure which provides a flow that, unlike the simple restrictor, is independent of the outlet, or load, pressure. This valve is a closed loop control in that the pressure drop is fed back onto the spool to control its position. However, the flow itself is controlled by the pressure supply control system. A typical pressure compensated valve is shown in Figure 6 the variable restriction being set by the control screw on the top of the valve.

The control method employed in the pressure compensated valve is used extensively in many other types of valve for a variety of hydraulic circuit functions as discussed in chapter 7.

5. Central bypass valve

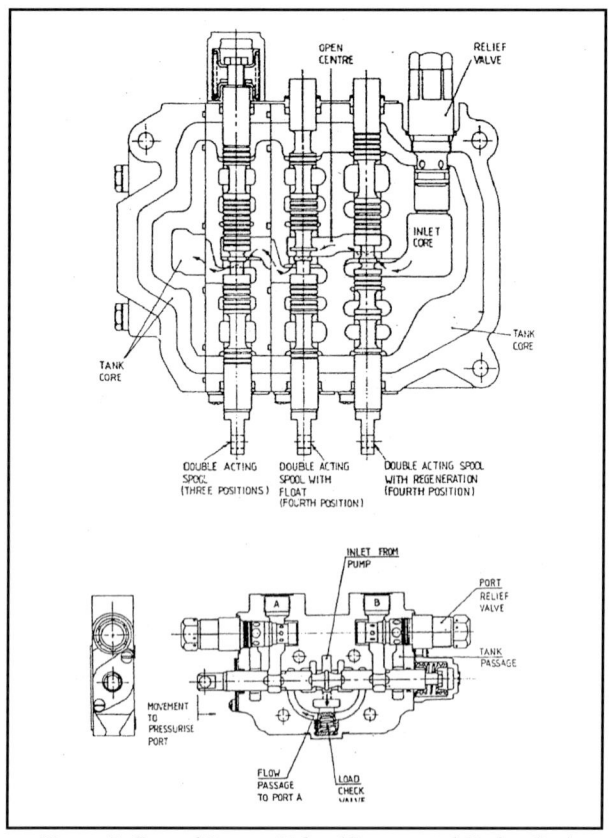

Figure 7. Central Bypass Valve (Commercial Hydraulics)

Central bypass valves provide a combination of directional control and restrictive metering where the metered flow is bypassed directly to the tank, or return line.

Figure 7 shows the construction of a typical bypass valve having three spools for operating three functions from a single pump. With the spools in the neutral position the pump flow passes through the open centres and as a spool is displaced it creates an increasing restriction to the pump flow thus raising the pressure.

Progressive displacement of the spool eventually opens a service port to the pump pressure but flow will only pass to this service providing the pump pressure is high enough to work against a loaded actuator that is connected to the outlet. Figure 8 shows the valve disposed from the centre position so that the 'A' port is connected to the supply and the 'B' port to return.

The load check valve is fitted so that it will open when the pump pressure is slightly higher than that required to move the actuator, or motor connected to either of the outlet ports. A number of circuit configurations are available with this valve some of which are described in chapter 7 on circuit design.

The spool movement also connects the other service port to the return line so that four-way control of actuators is provided. The major advantages of this valve are its simple construction and the generation of pump pressures that are only slightly higher than the maximum outlet pressure so minimising energy losses. The main disadvantages are the flow sensitivity with load pressure and the interaction that arises when two or more functions are operated simultaneously, the level of which varies with the relative pressures of the valve outlets.

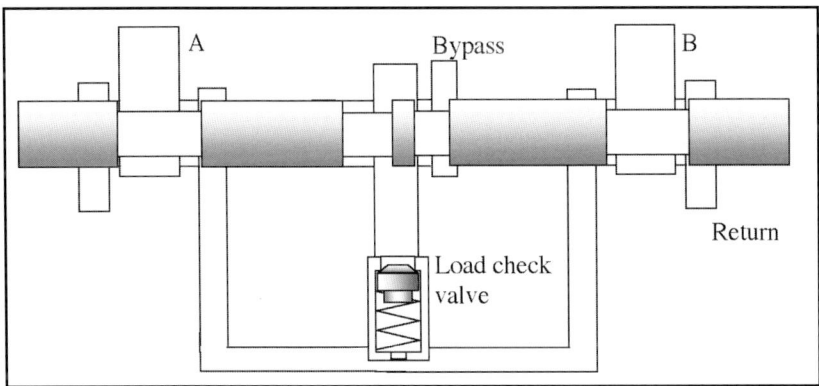

Figure 8. Bypass valve connecting the pump flow to port A and return to port B

CHAPTER SIX

ANCILLARY EQUIPMENT

6. ANCILLARY EQUIPMENT

Summary

Ancillary equipment basically includes those components or subsystems that are not directly involved in the major functions of the circuit. In many applications accumulators provide a supplementary flow source that can be used to meet high transient flow demands, compensation for leakage and absorption for pulsation and shock situations. These employ a volume of pressurised gas, usually nitrogen, which can be used to displace a fixed volume of hydraulic fluid as and when required.

All circuits require the fluid to be filtered, as contaminant particles in the fluid are the biggest cause of unreliability and failure of components and systems. These particles enter the system from the environment and are also generated by the wear process such as occurs in pumps and motors.

Inefficiencies in pumps, motors and actuators generate heat, which is absorbed by the fluid. It is necessary to be able to estimate the rate of such heat generation in order to install a cooler for the fluid if necessary. The fluid reservoir needs to be designed so as to enable absorbed air to be released and minimise the possibility of contaminant particles re-entering the system.

1. Introduction

This chapter is concerned with describing the function and relevant performance aspects of the following components:

- Accumulators

- Filters
- Coolers
- Reservoirs

2. Accumulators

2.1 Types

Accumulators are widely used in fluid power systems as a means of storing energy. Although weight and spring loaded types are sometimes used, those employing a pressurised gas are preferred because of their compactness and superior performance.

There are two main types of pressurised gas filled accumulators these using piston and collapsible bladders.

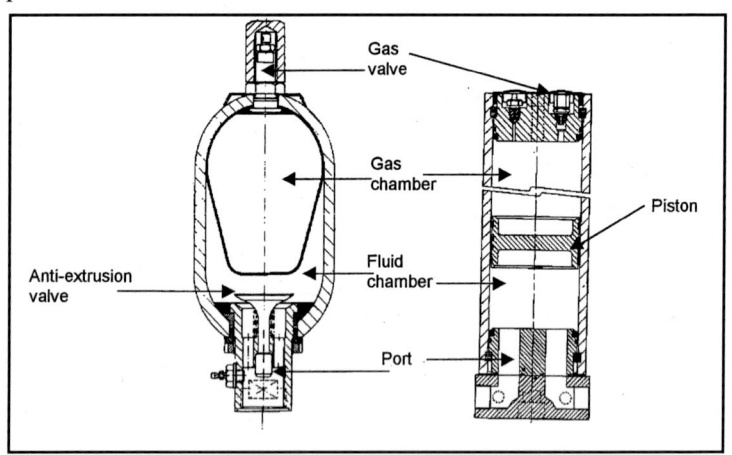

(a) Bladder type (Fawcett Christie) *(b) Typical piston type*
Figure1. Accumulators

As shown in Figure 1 the fluid is separated from the pressurised gas by either the bladder, Figure 1(a), or by a piston, Figure 1(b). The gas is usually nitrogen that is supplied via the gas valve with a pre-charge pressure that is determined by the pressure range required by the application.

Ancillary Equipment

Sealing of the piston is obviously important and there can be friction between the piston and the cylinder that can affect the liquid pressure level. This problem does not arise with the bladder types and extra gas can be added by the use of separate storage gas containers.

Accumulators are typically used for:

i) The supplementation of pump flow to meet high transient flow demands.
ii) Emergency supply.
iii) Leakage compensation.
iv) Shock alleviation.
v) Compensation required for volume changes due to temperature or pressure.
vi) Simple suspension elements.
vii) Pulsation absorption.

Legislation on the use of gas filled vessels requires certain maintenance procedures to be carried out which are described in the BFPA document P54 entitled: *Guide to Pressurised and Transportable Gas Containers Regulations and their Application to Gas Loaded Accumulators.*

2.2 Performance

The accumulator is initially charged to a pressure P_0 that is set at a level lower than the minimum operating pressure P_1. The pressure of the gas will vary with changes in the volume, but the relationship between these parameters will depend on the amount of heat transferred to the surroundings. It is usual to assume a polytropic expansion index for the gas, the value of which depends on the operating times and the duty cycle.

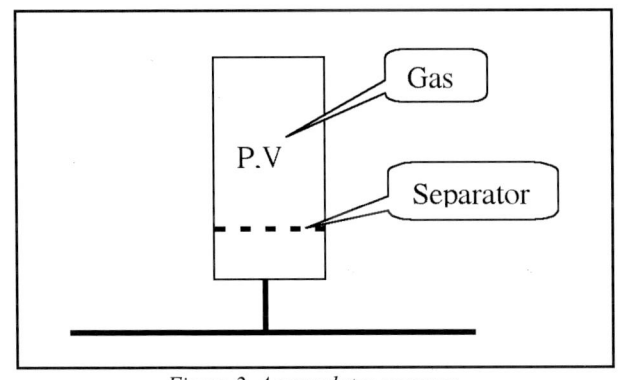

Figure 2. Accumulator pressure

Referring to Figure 2, the gas states are defined as:

Pre-charge: pressure P_0 and volume V_0.
Minimum operating: pressure P_1 and volume V_1.
Maximum operating: pressure P_2 and volume V_2.

For a gas having a mass 'm', an absolute temperature 'T' and a polytropic index 'n', the universal gas laws for a perfect gas give:

$$PV^n = constant \qquad (1)$$

$$PV = mRT \qquad (2)$$

R = Universal gas constant

The accumulator is connected to an appropriate point in the hydraulic system such that when the pressure falls the gas will expand and deliver a volume of fluid ΔV into the hydraulic system thus maintaining its pressure. The maximum volume is given by:

$$\Delta V = V_1 - V_2 \qquad (3)$$

Using a polytropic index, n_1, for compression from V_0 to V_2, for the period when the fluid pressure increases to its maximum value, equation (1) gives:

$$P_0 V_0^{n_1} = P_2 V_2^{n_1}$$

Thus

$$V_2 = \left(\frac{P_0}{P_2}\right)^{1/n_1} V_0 \qquad (4)$$

Also, for the gas expansion from V_2 to V_1 with a polytropic index of n_2:

$$V_1 = \left(\frac{P_2}{P_1}\right)^{1/n_2} V_2 \qquad (5)$$

Equations 4 and 5 with equation 3 give:

$$\Delta V = V_0 \left(\frac{P_0}{P_2}\right)^{1/n_1} \left\{ \left(\frac{P_2}{P_1}\right)^{1/n_2} - 1 \right\} \qquad (6)$$

Ancillary Equipment

The values of the polytropic indices cannot be accurately predicted and it is usual to take the value of n_1 as 1 (isothermal) and n_2 as γ for the gas, where this value is obtained for the expected operating temperature and pressure.

Thus equation 6 produces:

$$V_0 = \frac{\left(\frac{P_2}{P_0}\right)}{\left\{\left(\frac{P_2}{P_1}\right)^{1/\gamma} - 1\right\}} \Delta V \qquad (7)$$

This equation gives a conservative value for most applications. In certain cases, e.g. high or low temperatures, it may be necessary to apply a correction factor and, in those situations, information should be obtained from the manufacturer.

The value for γ can be obtained from Figure 3 that applies to real gases and should be used in accumulator sizing calculations.

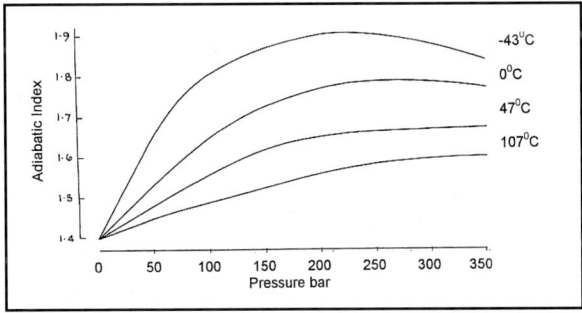

Figure 3. Variation in adiabatic index with pressure and temperature for nitrogen.

3. Contamination control

3.1 Components

The selection of inadequate filters or poor maintenance procedures can cause excessive contamination levels that may result in the unreliable operation and breakdown of hydraulic components. Filtration systems should, therefore, be designed such that the fluid cleanliness level is better than that specified by the component manufacturers.

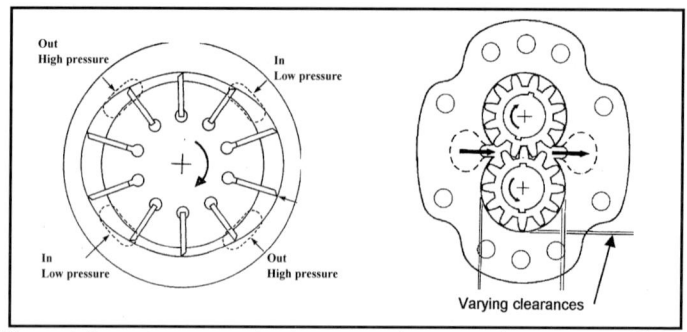

Figure 4. Important contamination aspects in vane and gear pumps (Eaton)

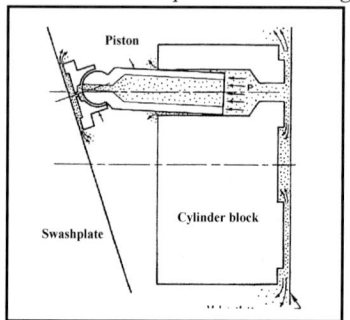

Figure 5. Wear particle generation in piston pumps (Eaton)

Figures 4 and 5 indicate where metal particles are generated in pumps and also where particles in the incoming fluid will accelerate the wear process. The clearances in pumps, motors and valves are of the order of a few microns and it is, therefore, essential that the fluid be kept clean at this level.

3.2 Filters

The main features of a replaceable element high-pressure filter are shown in Figure 6. As the fluid passes radially inward through the element contaminant is trapped in the material. With time the pressure drop across the filter will increase at a rate that is dependent on the fluid condition and eventually this will cause the bypass valve to open thus passing contaminated fluid directly into the system. However, the pressure drop can be monitored either mechanically or by electronic methods and this aspect is an important feature in a properly maintained system.

Ancillary Equipment

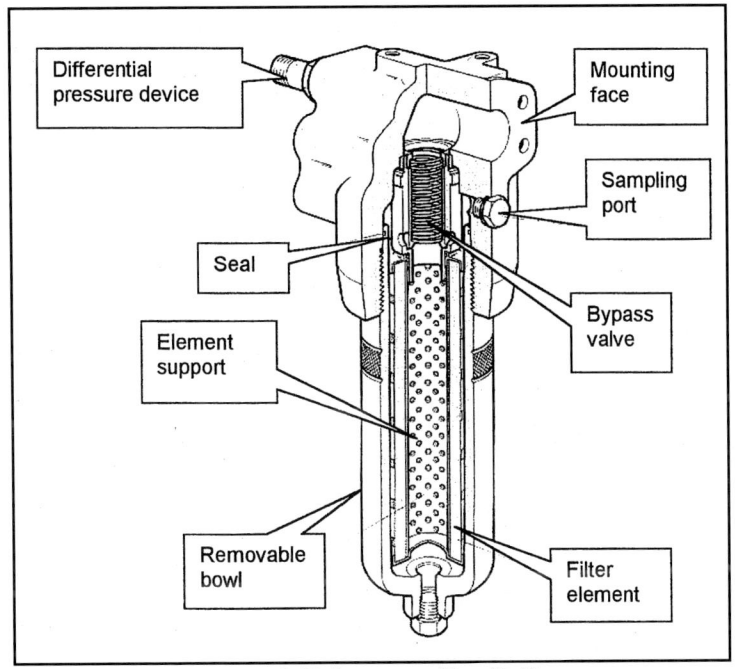

Figure 6. High-pressure filter (Pall)

A major problem associated with filtration is that its effect cannot be seen because of the small size of particles that can cause poor system reliability and component failure so it is important that monitoring of the filter condition is carried out on a regular basis. Sampling techniques and the measurement of the contaminant concentration provide an improved basis for monitoring the condition of the hydraulic system.

The performance of a filter is based on its ability to trap particles which is defined by its beta ratio, β, that is obtained from appropriate test methods.

The beta ratio is defined as:

$$\beta_x = \frac{No.\ of\ particles > particlesize\ 'x'\ upstream}{No.\ of\ particles > particlesize\ 'x'\ downstream}$$

The beta ratio is defined for particle sizes above the given level because the number of trapped particles varies with the size, which is referred to as a distribution.

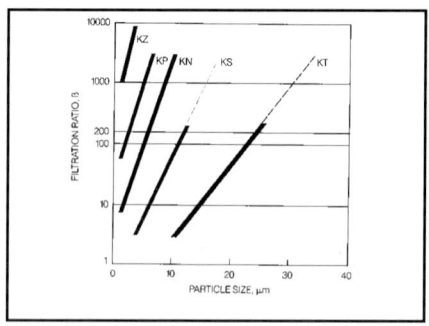

Figure 7. Beta ratio for filters (Pall)

Filters are selected on the basis of achieving the desired contamination levels and having sufficient contaminant holding capacity to maintain the required contamination levels under the worst envisaged circumstances. Various selection methods are available from different filter manufacturers, the majority of which are based on an absolute filter rating at a given β ratio. Figure 7 contains an example showing the performance of different elements with a $β_x$ = 200 rating where x is the minimum particle size for a beta ratio of 200.

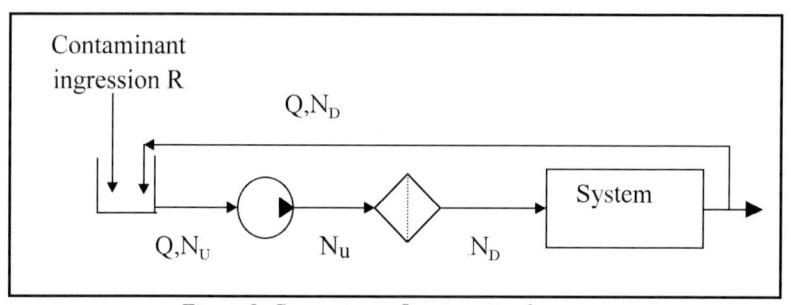

Figure 8. Contaminant flow in a simple system

Generally it is not feasible to analyse systems in respect of the generation of contaminant particles and ingression from the environment. However a simple model such as that in Figure 8 can be used to show that:

$$N_U = \frac{R}{Q} \frac{\beta}{\beta - 1} \text{ and}$$

$$N_D = \frac{R}{Q(\beta - 1)}$$

Ancillary Equipment 71

For beta ratios in the region of 10 and above, N_D reduces as the inverse proportion of the beta ratio which is represented by the chart in Figure 9.

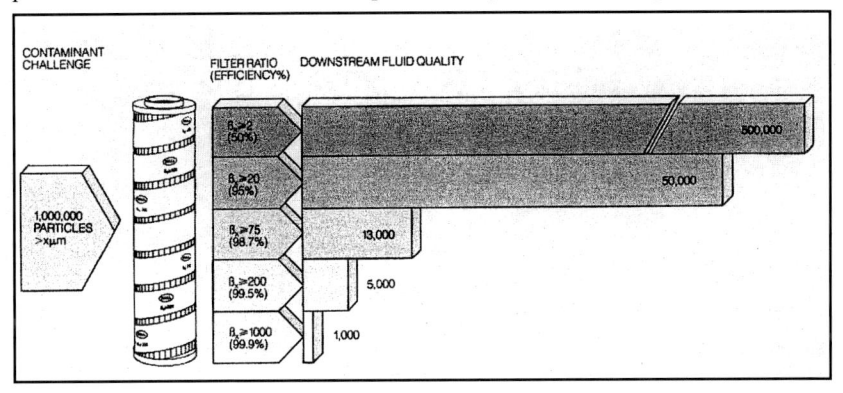

Figure 9. Contaminant levels and the beta ratio (Pall)

The procedure described in the BFPA document, P5, contains sufficient information for the selection of an appropriate filter in a given installation. Contaminant levels are denoted by an ISO code that is related to the numbers of particles of sizes greater than 4, 6 and 14 microns respectively. This is shown in Figure 10.

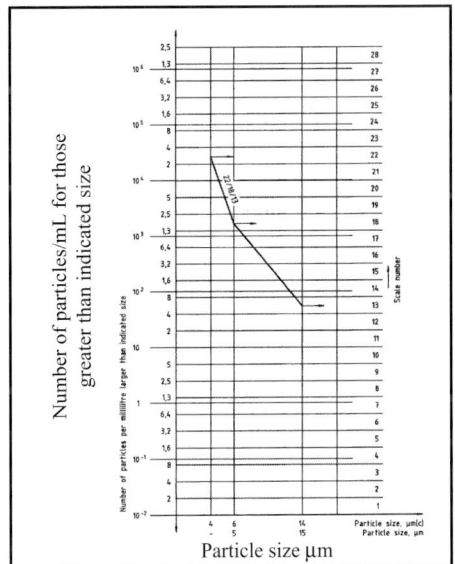

Figure 10. ISO 4406 standard code for contamination levels.

4. Coolers

Heat is generated in the fluid in hydraulic systems because of pressure losses in pipes, fittings and, particularly, in control valves where the rate of heat dissipation can be of the same order of the power being produced at the system output. The temperature created in the fluid will depend on the system duty cycle and its environment, as natural heat convection from pipes and reservoirs is not at a very significant level.

In industrial systems the fluid temperature is usually around 50 to 60^{0}C and in mobile equipment this can be as high as 80^{0}C. The condition of most hydraulic oils is significantly affected by operation at high temperatures, which will shorten the life of the oil and reduce its viscosity to unacceptable levels.

Hydraulic component manufacturers specify the viscosity range to be used and in the main these will call for a minimum value of around 20 cSt although some will operate satisfactorily at 10 cSt and less. It is important to realise that the volumetric efficiency of pumps and motors is significantly affected by operation with low viscosity fluids, which will cause a further increase in the heat load.

It is therefore necessary to estimate the amount of heat generation in order to establish the size of cooler that will be required to maintain a satisfactory fluid temperature.

4.1 Cooler types

Coolers use either air or water as the cooling fluid. In water coolers the water flows through the tubes and the oil across the tubes, the latter guided in its flow path through the shell by baffle plates. There are two common constructions; in the first the tubes are arranged in a U-bundle with a single tube sheet, in the second two tube-sheets are used in a straight tubing arrangement.

The maximum oil pressure that the cooler can be subjected to is limited by the shell, a typical figure would lie in the range 15 to 30 bar. The pressure drop associated with the oil flow through the cooler is usually small, of the order of 1 bar. Water coolers are more compact than those using air and, providing an adequate supply of cool water is available, they are less sensitive to environmental conditions. In some cases it may be necessary to fit a strainer at the cooler inlet in order to prevent blockage of the water flow.

Air coolers using fans to create the necessary airflow are of a lighter construction than water coolers but are larger and sensitive to the environmental conditions, which needs to be considered. Air coolers are mostly used for

Ancillary Equipment

mobile applications and usually can only work with oil pressures up to around 7 bar.

For both types of coolers automatic temperature controls are available using thermostats to either control the water inlet flow or the speed of the fan.

4.2 Thermodynamic aspects

$$C_P \Delta T \frac{dm}{dt} = Q \Delta P$$

as $\frac{dm}{dt} = \rho Q$, then $\Delta T = \frac{\Delta P}{\rho C_P}$

Where $\frac{dm}{dt}$ = Mass flow rate   kg s^{-1}

C_p = Specific heat   2.1 kJ kg^{-1} K^{-1} (typical value)
T = Temperature   K
ρ = Density   870 kg m^{-3} (typical value)
ΔP = Pressure loss   N m^{-2}
Q = Flow   m^{3}s^{-1}

Therefore, for a system with a 100 bar pressure loss, the temperature rise will be:

$$\Delta T = \frac{\Delta P}{\rho C_p} = 5.5°C$$

The input power,

$$W_0 = Q \Delta P \quad \therefore \Delta T = \frac{W_o}{\rho C_p Q}$$

where W_o is the power input given in Watts.

4.3 Cooler characteristics

The cooling characteristics are usually presented in the form shown in Figure 11, which will apply for a particular value of inlet temperature difference. For different values a correction factor is applied to suit the application.

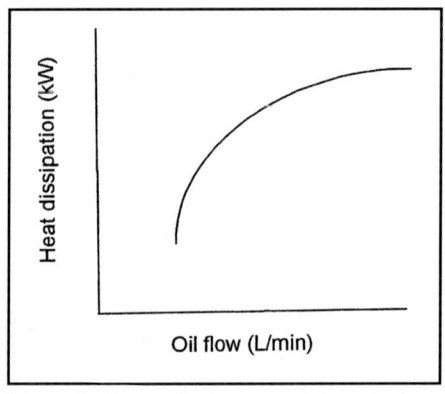

Figure 11 Cooler Performance Characteristics

5. Reservoirs

If possible, the reservoir design should be such that any entrained air is released before the fluid is passed to the system inlet. A baffle can be used to increase the flow circulation and hence improve the air release. It also reduces the fluid movement due to motion of the reservoir itself.

Passing the return flow through a diffuser in the reservoir reduces the fluid velocity and, by directing the fluid away from the bottom of the reservoir, reduces the re-entrainment of solid contaminant and water from the bottom of the reservoir. Transient changes in the fluid level and the release of air require the fitment of a breather. This must contain a filter that is sized to the minimum requirements of the system. It is preferable that the fluid is filtered during topping-up or replenishing.

CHAPTER SEVEN

CIRCUIT DESIGN

7. CIRCUIT DESIGN

Summary

The selection of hydraulic components for use in a given application is determined by their ability to meet the required specification within the desired cost framework. A variety of components can be arranged to fulfil a given function by using different circuit configurations as the fluid power system designer has the freedom, within the constraints set by the preferences of the machine builder and /or the user, to select components of his choice.

This freedom makes it difficult to summarise circuit design however, the designer needs to be able to justify the circuit on the basis of technical considerations. This chapter therefore describes and, where applicable, evaluates a variety of circuit options that can be used for the range of functions generally encountered in the application of fluid power systems.

1. Introduction

To a very large degree the main function of hydraulic circuits is to control the flow to one or several actuators as required by the application. There are, however, a variety of methods for controlling flow, some of which act indirectly by using pressure as the controlling parameter.

The circuits discussed in this chapter include :
- Directional control and valve configurations.
- Velocity controls with constant supply pressure.
- Velocity controls with load sensing.

- Variable displacement pump controls.
- Hydrostatic transmissions.
- Load control.
- Contamination control

2. Pressure and Flow

Hydraulic systems provide flow from the pump that is directed to one or more actuators (motors) at a pressure level that satisfies the highest demand. Where a single output is being driven the pump pressure will float to the level demanded by the load. However, even for such simple systems the method that is employed to provide variable flow needs to be evaluated in order to ensure that best efficiency is obtained. In circuits with multiple outputs this aspect can be more difficult to evaluate.

For operation at pressures and flows that are lower than the required maximum values the efficiency of the system will depend on the type of pump being used (i.e. fixed or variable displacement). This can be represented diagrammatically as in Figure 1.

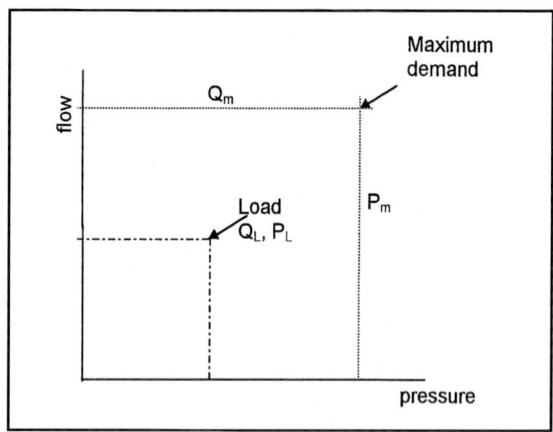

Figure 1. Flow and pressure variation

For fixed displacement pump systems it is clear from Figure 1 that excess pump flow will have to be returned to the reservoir so that the power required by the pump is greater than that being supplied to the load. The level of inefficiency incurred is dependent on the ratio between the pressure required by the load and that at the pump outlet which can be controlled at the maximum level by the relief valve or at lower pressures by various types of bypass valves.

For variable displacement pumps, the generation of excess flow can be avoided. However, the level of pump pressure will depend on the method that is used for controlling the displacement but clearly there is scope for achieving much higher efficiencies than with fixed displacement pumps.

Each of these control methods will require a particular circuit design employing components that have been described in the previous chapters.

3. Directional control

Valves used for controlling the direction of the flow can be put into fixed positions for this purpose but many types are frequently used in a continuously variable mode where they introduce a restriction into the flow path.

3.1 Two position valves

A four-way valve with two positions for changing direction of the flow to and from an actuator is shown in Figure 2. For supply flow, Q, the actuator velocities will be:

$$\text{Extend } U_E = \frac{Q}{A_P}; \quad \text{Retract } U_R = \frac{Q}{A_A}$$

Here, the actuator areas are A_P for the piston and A_A for the annulus or rod end of the actuator. Hence,

$$U_R > U_E \text{ as } A_P > A_R$$

Any external forces (F) that are acting on the actuator rod must be in opposition to the direction of motion. For reversing force applications it will be necessary to apply restrictor control which will be discussed later in the chapter. These forces will create a supply pressure that is

$$= \frac{F}{A_P} \text{ or } \frac{F}{A_A}$$

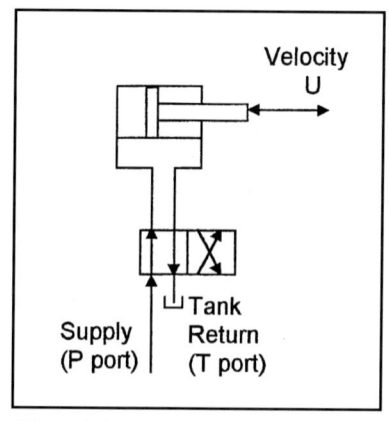

Figure 2. Two position four-way valve

Three-way valves are used in applications where only one side of the actuator needs a connection from the supply. A typical example for this is the operation of the lift mechanism on a fork lift truck, as shown in Figure 3 where the actuator is lowered under the action of the weight.

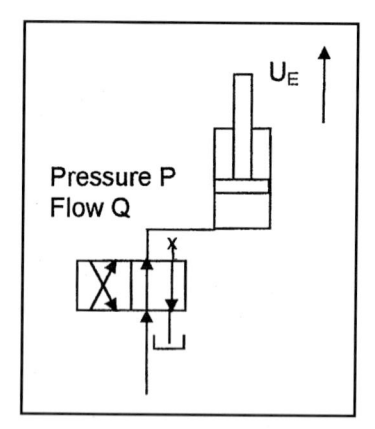

Figure 3. Two position three-way valve

3.2 Three position valves

Three position valves have a third, central position that can be connected in different configurations. These variants are described.

Circuit Design

Closed Centre Valves (Figure 4)

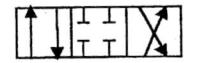

Figure 4.

Closed centre valves block all of the four ports. This prevents the actuator from moving under the action of any forces on the actuator. The supply flow port is also blocked which may require some means of limiting the supply pressure unless other valves are being supplied from the same source. The limitation of the supply pressure can be made by appropriate pump controls or by a relief valve.

Tandem Centre Valves (Figure 5)

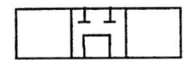

Figure 5.

Tandem centre valves block the actuator ports but the supply is returned to the tank at low pressure. If other valves are being supplied from the same source this type of valve may not be used – unless connected in series.

Open Centre Valves (Figure 6)

Figure 6.

Open centre valves connect all of the four ports to the tank so that the supply and the actuator pressure are at low pressure. This allows the actuator to be free to move under the action of any external forces.

Figure 7.

Where it is necessary to block the supply flow the configuration shown in Figure 7 can be used.

4. Load holding valves

The radial clearance between the valve and its housing of spool valves is carefully controlled in the manufacturing process to levels of around 2 micron. The leakage through this space, even at high pressures, is small but for applications where it is essential that the actuator remains in the selected position for long periods of time (e.g. crane jibs where any movement would be unacceptable) valves having metal-to-metal contact have to be used.

Check valves usually employ metal-to-metal contact but they are only open in one direction under the action of the flow into the valve. For their use in actuator circuits it is necessary that they are open in both directions as required by the DCV. This function can be obtained from a Pilot Operated Check Valve that uses a control pressure to open the valve against reverse flow.

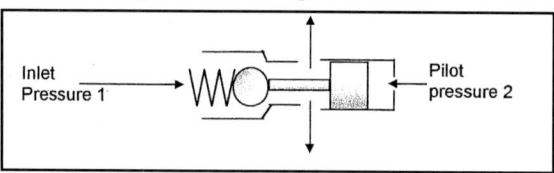

Figure 8. Pilot operated Check Valve

Figure 8 shows a typical pilot operated check valve (POCV) whereby a pilot pressure is applied onto the piston to force open the ball check valve to allow flow to pass from port 1 to port 2 when the check valve would normally be closed. The ratio of the piston and valve seat areas has to be chosen so that the available pilot pressure can provide sufficient force to open the valve against the pressure on port 1.

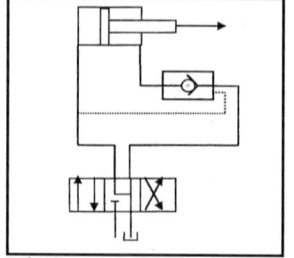

Figure 9. Actuator Circuit using a POCV

The use of a POCV is shown in Figure 9 where the external force on the actuator is acting in the extend direction. With the DCV in the centre position the check valve will be closed because the pilot is connected to the tank return line that is at low pressure. Opening the DCV so as to extend the actuator causes the piston side pressure, now connected to the supply, to increase.

When this pressure reaches the level at which the check valve is opened against the pressure generated on the rod side of the actuator by the load force, the actuator will extend. The ratio of the pilot and ball seat diameters needs to be such that the pressure areas cause the POCV to be fully open against the annulus pressure. If the pilot pressure is insufficient to open the valve because of an intensified pressure at the check valve inlet from the actuator annulus and/or back pressure on the POCV outlet due to restriction in the DCV, oscillatory motion can result.

5. Velocity control

The velocity of actuators can be controlled by using a number of different methods. In principle the various methods can be employed for both linear and rotary actuators or motors but in some cases it may be necessary to refer to the manufacturer's literature for guidance.

5.1 Meter-in control

Meter-in control refers to the use of a flow control at the inlet to an actuator for use with actuators against which the load is in opposition to the direction of movement.

For a meter-in circuit that uses a simple adjustable restrictor valve selection of the DCV to create extension of the actuator will cause flow to pass through the restrictor into the piston end of the actuator. The required piston pressure, P_p, will depend on the opposing force on the actuator rod. With a fixed displacement pump delivering a constant flow, excess flow from the pump will be returned to tank by the relief valve at its set pressure, P_{Smax}. Consequently, the available pressure drop, $P_{Smax} - P_p$ will determine the flow delivered to the actuator for a given restrictor opening.

With this system the flow, and hence the actuator velocity, will vary with the load force. For systems where such velocity variations are undesirable a pressure compensated flow control valve (PCFCV) can be used. This valve will maintain a constant delivery flow providing that the pressure drop is greater than its minimum controlled level that is usually in the region of 10 to 15 bar.

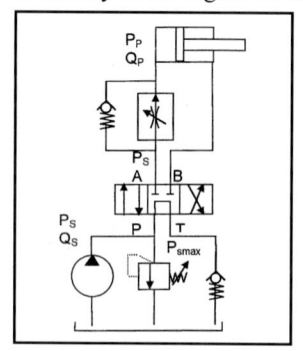

Figure 10. Meter-in Control for Actuator Extension

Figure 10 shows a typical system in which the flow control is bypassed with a check valve for reverse operation of the actuator. If the load force varies considerably during operation, there will be transient changes in actuator velocity at a level that depends on the mass of the load.

For example, when the load force suddenly reduces, the piston pressure will reduce but at a rate that is dependent on the fluid volume and its compressibility and the mass of the load. During the period that the pressure is greater than the required new value, the actuator will accelerate and, as it does so the piston pressure will fall. The pressure can then fall below the new level and deceleration results and damped oscillations can occur as shown in Figure 11.

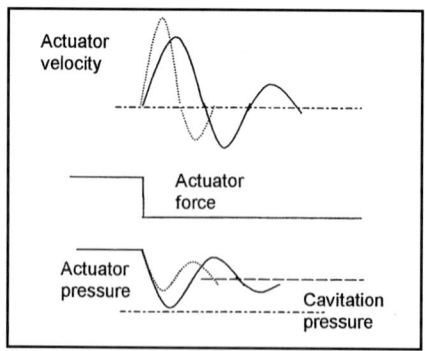

Figure 11. Pressure and Velocity Variations with Meter-In Control

In some situations the mass of the load can be such as to cause problems of cavitation and overrunning because the pressure falls transiently to a level at which absorbed air is released. If the pressure falls low enough the fluid will vaporise. Both of these phenomena are referred to as cavitation and noisy operation, and damage to the components can be the result.

The changes represented by dotted lines in Figure 11 are for a low inertia load that creates a lower magnitude of the pressure oscillations and, hence reduces the possibility of cavitation.

A check valve having a spring cracking pressure that is high enough to suppress cavitation is sometimes used as shown in Figure 11 but this has the disadvantage of increasing the pump pressure and thus reducing the efficiency and increasing the heating effect on the fluid.

5.2 Meter-out control

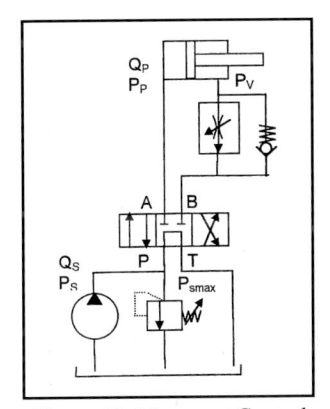

Figure 12. Meter-out Control

For overrunning load forces and/or those with a large mass, meter-out control is used where the actuator outlet flow during its extension passes through the restrictor or PCFCV as shown in the circuit of Figure 12.

The flow control operates by controlling the actuator outlet pressure at the level required to oppose the forces exerted on the actuator by the load and by the piston pressure which is the same as that of the pump. This prevents cavitation from occurring during transient changes arising from load force variations or due to forces that act in the same direction as the movement (i.e. pulling forces).

This system can, however, cause high annulus pressures to occur from the intensification of the piston pressure together with the pressure created by pulling forces. Further, when compared to meter-in, the rod and piston seals have to be capable of withstanding high pressures that may require a higher cost actuator to be used.

5.3 Bleed-off control

For the fixed displacement pump system shown in Figure 13, excess flow is bled off from the supply so that the pump pressure is now at the same level as that required at the actuator piston.

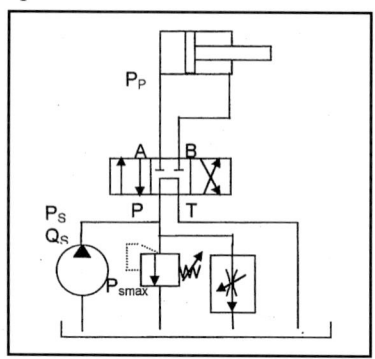

Figure 13. Bleed-off Control

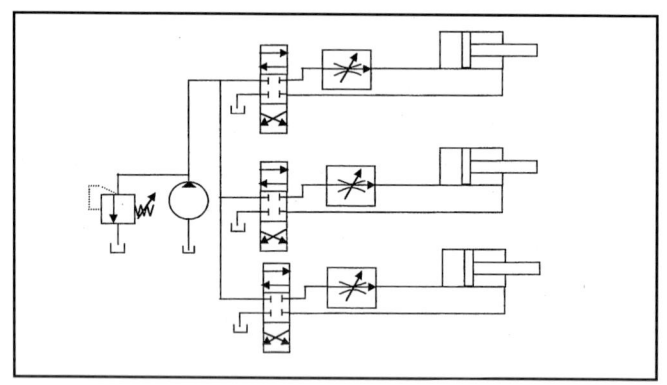

Figure 14. Multiple Actuator Circuit with Meter-in Control

Bleed-off control is therefore more efficient than meter-in and meter-out because of the lower pump pressure. However, as for meter-in, it cannot be used with pulling loads and it can also only be used to control one actuator at a time from the pump. This is in contrast to meter-in and meter-out where several actuators can be supplied by a single pump as shown in Figure 14.

Meter-in and meter-out controls can be supplied from a variable displacement pump that is operated with a constant pressure control (pressure compensated) which reduces the power wastage that is inherent with a fixed displace-

ment pump. This is demonstrated by making a comparison of the efficiencies as follows:
For meter-in control the power efficiency,

$$\eta = \frac{Q_P P_P}{Q_S P_S}$$

For a pressure compensated pump the power efficiency,

$$\eta = \frac{Q_P P_P}{Q_P P_S} = \frac{P_P}{P_S} \text{ as } Q_S = Q_P.$$

Thus referring to Figure 1, the pump flow is always equal to that of the load, Q_L. The pump is still capable of achieving the maximum demand, which is referred to as the 'corner power' of the pump. The fixed displacement pump operates at this rating continuously because of the use of the relief valve to control the flow to the actuator.

The flow control methods described in this section are usually preset in a system that is being used on a continuous basis such as for a production machine (e.g. injection moulding) where possibly the operations are being carried out sequentially. It would normally be expected that the duration of, say, actuator movement is small in relation to the overall cycle time so that the power losses are relatively small. Where a continuously variable flow control is required alternative components and circuits need to be considered.

5.4 Four way valve restrictive control

Four way valves can be used to control the velocity of actuators by introducing both meter-in and meter-out restrictions into the flow path as shown in Figure 15.

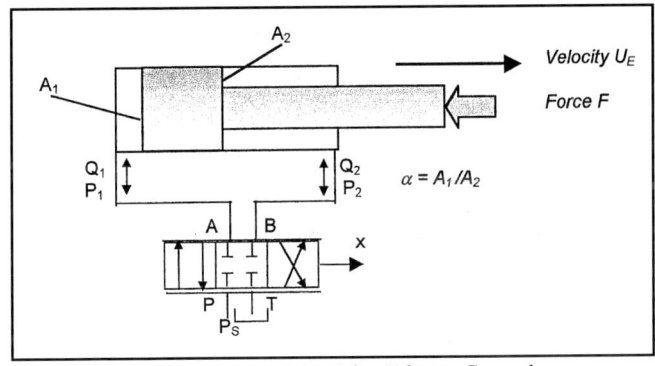

Figure 15. Four Way Valve Velocity Control

The valve position is fully variable and can be controlled by:

- Direct lever manual input.
- Hydraulic pilot operated from an input lever or joystick.
- Proportional solenoid

For proper design of the circuit and appropriate component selection it is necessary to analyse the system in order to determine the actuator pressures as a function of the actuator load force. In this system the pressure drop across each valve land needs to be considered as a function of the flow and the valve position or opening.

5.4.1 Analysis of the valve/actuator system

5.4.1.1 Actuator extending

<u>Valve flow characteristics</u>

For the parameters shown in Figure 15 the valve flows are given by:

$$Q_1 = K_1 \times \sqrt{P_s - P_1}$$

$$K_1 = R_1 \sqrt{\frac{2}{\rho}}$$

where R_1 = effective metering area
(e.g., $= C_Q \pi d$ for annular ports)
For zero pressure in the return line:

$$Q_2 = K_2 \times \sqrt{P_2}$$

$$K_2 = R_2 \sqrt{\frac{2}{\rho}}$$

where R_2 = effective metering shape parameter
(this may be different to R_1 in some values)

Circuit Design

Also,

$$U_E = \frac{Q_1}{A_1} = \frac{Q_2}{A_2}$$

These equations give:

$$P_S - P_1 = (\frac{A_1}{A_2}\frac{K_2}{K_1})^2 P_2 = (\frac{\alpha}{R_S})^2 P_2$$

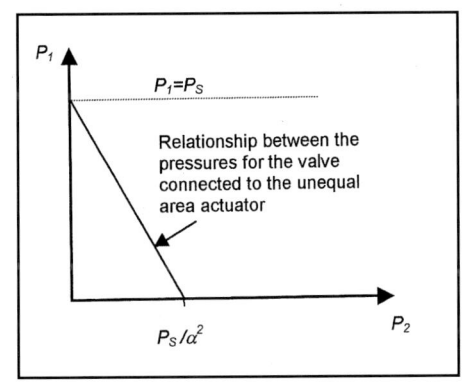

Figure 16. Valve Pressures during Extension

For a valve spool that has symmetrical metering, $R_S = 1$

Then
$$P_S - P_1 = \alpha^2 P_2 \qquad (1)$$

Equation 1 relates the pressures P_1 and P_2 for the valve connected to the unequal area actuator as represented Figure 16. The flow from the annulus is lower than that to the piston because of its smaller area, which results in a lower pressure drop in the valve.

Actuator force

$$P_1 A_1 - P_2 A_2 = F$$

$$P_1 = \frac{P_2}{\alpha} + \frac{F}{A_1} \qquad (2)$$

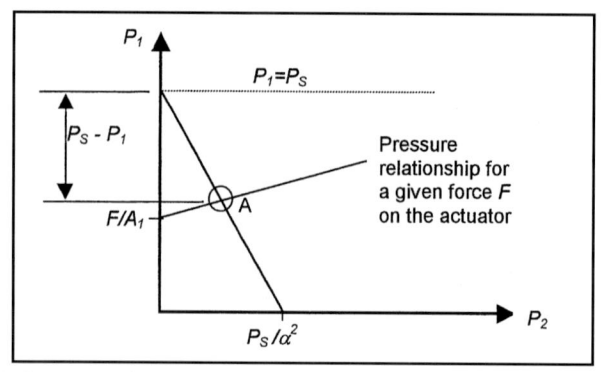

Figure 17. Interaction between the flow and the force characteristics during extension

Equation 2 relates the actuator pressures for a given actuator force with a positive force that acts in the opposing direction to the extending movement of the actuator. This relationship is shown on Figure 17 and the pressures for the valve controlling the actuator are determined by the intersection of the lines for the actuator and the valve at point A.

The values of these pressures can be obtained from equations 1 and 2 for the valve and actuator. The force acting on the actuator can be in either direction.

Using a force ratio defined by:

$$R = \frac{F}{P_s A_p}$$

we get:

$$\frac{P_1}{P_s} = \frac{(1+R\alpha^3)}{(1+\alpha^3)} \qquad (3)$$

and

$$\frac{P_2}{P_s} = \frac{\alpha(1-R)}{(1+\alpha^3)} \qquad (4)$$

Variations of the force will change the position of the line on Figure 17 that represents the actuator and, hence, change the position of the point of intersection, A, that will change the pressure levels according to Equations 3 and 4.

For a value of $R = 1$, the force will be the maximum available and is the stall force for the system. For a given application the actuator area is chosen to provide the desired stall force at the chosen supply pressure. This will then determine the actuator pressures that are obtained for any other value of the force.

Circuit Design

5.4.1.2 Actuator retracting

For retraction of the actuator the valve position is reversed, the annulus now being connected to the supply and the piston to the return line. Following the same method for a symmetrical valve ($K_1 = K_2$) as for the actuator extension we get:

$$Q_2 = K_1 x \sqrt{(P_S - P_2)}$$

and

$$Q_1 = K_1 x \sqrt{P_1}$$

and as

$$Q_1 = \alpha Q_2$$

Then

$$P_2 = P_S - \frac{P_1}{\alpha^2} \qquad (5)$$

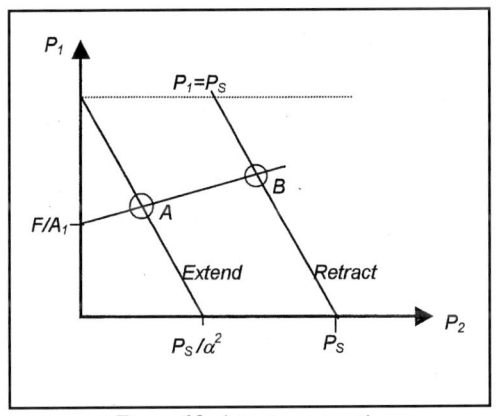

Figure 18. Actuator retracting

This relationship between P_1 and P_2 for the retracting actuator is shown in Figure 18 where point B is the operating condition for retraction of the actuator. It can be seen that when the valve is reversed there is a pressure change between points A and B.

The equations for the pressures for actuator retraction are:

$$\frac{P_2}{P_s} = \frac{\alpha^3 - \alpha R}{1 + \alpha^3} \qquad (5)$$

$$\frac{P_1}{P_s} = \frac{\alpha^2 (1+\alpha R)}{(1+\alpha^3)} \qquad (6)$$

The actuator velocity is determined from the flow equation for the valve using the appropriate values for the pressures P_1 and P_2. The selection of a valve having the necessary capacity is determined from the maximum required velocity condition.

It should be noted that neither of the pressures can be less than zero so, for example, during extension, the maximum value of the force ratio, R, that is permissible in order to avoid cavitation of the flow to the piston is given by:

$$\frac{P_1}{P_s} = 0 = \frac{1+R\alpha^3}{1+\alpha^3}$$

$$\therefore R \not> \frac{1}{\alpha^3} \text{ to prevent cavitation}$$

For which condition:

$$P_2 = P_s \alpha \frac{(1+1/\alpha^3)}{(1+\alpha^3)}$$

$$\therefore P_2 = \frac{P_s}{\alpha^2}$$

5.4.2 Valve sizing

The valve size needs to be selected such as to provide the flow required for the specified velocity and force conditions. In general, it is desirable to operate the system close to its condition of maximum efficiency, which can be obtained by considering the power transfer process.

For simplicity consider an equal area actuator for which the power transmitted to the load is given by;

$$\text{Power, } E = P_m Q_m$$

$$\therefore E = (P_{s\,max} - \Delta P_v) Q_m$$

Here, P_m is the load pressure difference, Q_m the actuator flow and ΔP_V is the valve pressure drop which, from the orifice equation, can be expressed as:

$$Q_m = C_q A \sqrt{\frac{2\Delta P_V}{\rho}}$$

$$\therefore \Delta P_V = \left(\frac{\rho}{2 C_q^2 A^2}\right) Q_m^2 = k Q_m^2$$

where A = the valve orifice area.

Hence:

$$E = (P_{s\,max} - k Q_m^2) Q_m$$

For maximum power at the load:

$$\frac{dE}{dQ_m} = 0 = P_{S\,max} - 3 Q_m^2$$

$$\therefore k Q_m^2 = P_m \quad \text{giving} \quad P_m = \frac{2}{3} P_{S\,max}$$

Based on this analysis an approximation is often applied from the maximum power condition to give the best combination of valve and actuator sizes for a given supply pressure such that the stall thrust is equal to two-thirds ($\frac{2}{3}$) of the force at maximum power.

The actuator size and the supply pressure can be selected to provide this stall thrust and the valve size is then determined on the basis of the maximum required value of

$$\frac{Q}{\sqrt{\Delta P_V}}$$

Valves are rated by determining the flow at a total fixed pressure drop across both ports and for a given valve position. Thus the flow at any other pressure drop is given by:

$$Q = Q_R \sqrt{\frac{\Delta P}{\Delta P_R}}$$

Where Q_R = rated flow and ΔP_R = rated pressure drop. ΔP_R is sometimes given as the pressure drop through only one of the metering lands.

5.4.3 Valves with non-symmetrical metering

The use of valves in which the metering is non-symmetrical, normally by machining metering notches of different shapes, provides two major advantages over symmetrical valves which are:

• The possibility to increase the maximum negative, or overrunning force during extension (meter-out control)
• The avoidance of the pressure change during reversal of the valve - as seen from Figure 19 the valve characteristic is a single line for both directions of movement.

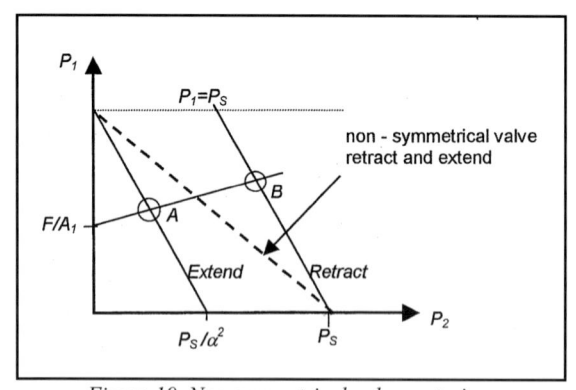

Figure 19. Non-symmetrical valve metering

The dotted line on Figure 19 shows the characteristics where the valve metering ratio, R_s, has the same value as the actuator area ratio, α.
Figure 20 shows the output velocity U plotted against the force ratio R in

Circuit Design

comparison with the particular case of an equal area ratio actuator for which $\alpha = 1$. The equal area actuator has a symmetrical characteristic and is often used in servo systems for this reason. The dotted line shows the effect of asymmetrical metering where the valve-metering ratio is the same as the area ratio of the actuator.

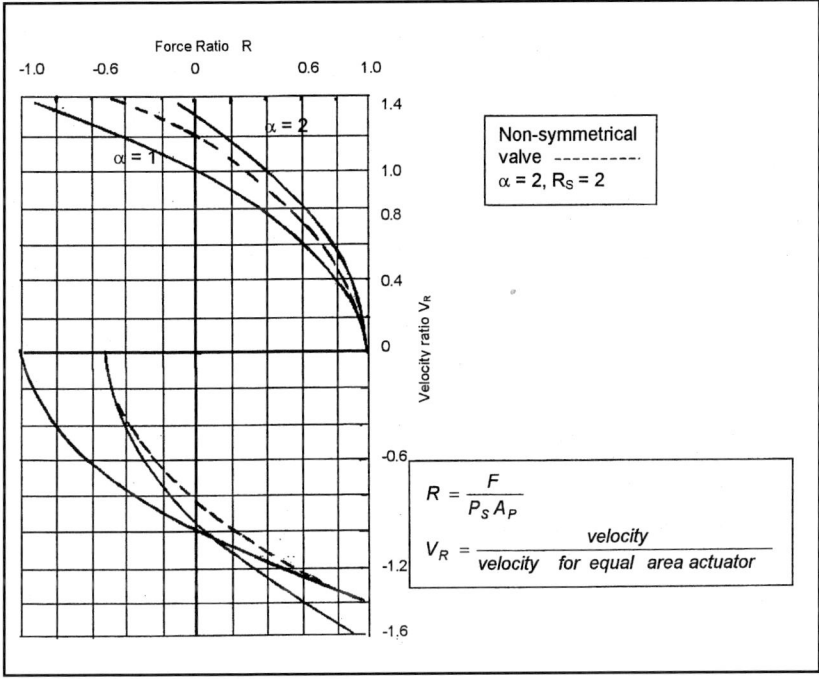

Figure 20. Load locus of velocity ratio against force ratio

5.5 Bypass control with fixed displacement pumps

5.5.1 Open centre valves

Open centre valves that use a central bypass combine the use of bleed off control with the directional control function. A circuit is shown in Figure 21 for two valves connected in series. Operation of a valve creates an outlet flow from the valve, the magnitude of which will depend on the valve opening and the required outlet pressure as shown graphically in Figure 22. When the valve is moved to an extreme position on either side, the central bypass will be closed and all of the pump flow passed to the outlet.

On starting a loaded actuator, when the valve is opened the pump pressure will rise to a level that is determined by the amount of opening. The load check valves are placed in the circuit to prevent the actuator reversing should the pump pressure be less than that required by the actuator. The relief valve is required to protect the pump and system from excessive pressures.

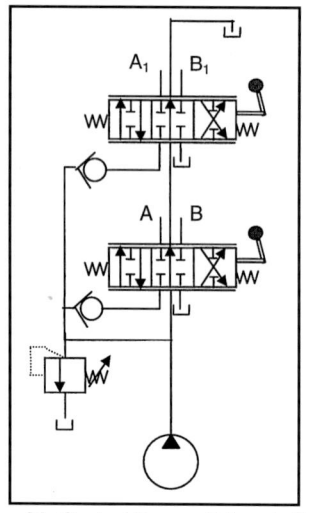

Figure 21. Central Bypass Valves in Series

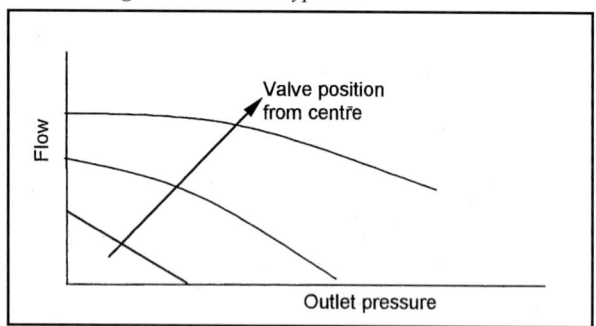

Figure 22. Bypass Valve Characteristics

The valve is load sensitive in that, for a given valve position, the flow reduces with increases in the outlet pressure. When both valves are operated simultaneously there will be interaction causing the flows to vary with changes in either of the outlet pressures.

Machining a notch in an overlapped valve land often forms the valve metering area. The typical valve configuration in Figure 23 for three valve spool positions shows this feature in the central bypass land. By using different numbers of notches and/or their sizes different metering characteristics can be obtained from the same size of valve spool and also differential metering related to the selected flow direction which includes meter-out control for negative loads.

Figure 23. Central Bypass Valve with Notched Metering Edges in 3 Positions

5.5.2 Closed centre valves with load sensing and pressure compensation

Closed centre valves can be used with fixed displacement pumps whereby excess flow is bypassed from the pump output under the action of a spring loaded valve that senses the pressure drop between the pump and the load pressure. This creates a pump pressure that is about 20 bar above that of the load that provides the efficiency benefits from bleed off control. This also provides pressure compensation and thus avoids the load sensitivity obtained from central bypass valves.

For use in circuits where several actuators are to be controlled valves are available in which load sensing can be incorporated to operate the bypass valve using a pilot signal from the highest load pressure.

Such a circuit is shown in Figure 24 in which individual pressure compensators are included to maintain constant flow through the valve that is supplying the actuator that is at the lowest pressure. The highest load pressure is sensed by the check valves and passed to the bypass-regulating valve, the opening of which is such as to deliver the flow required by both valves.

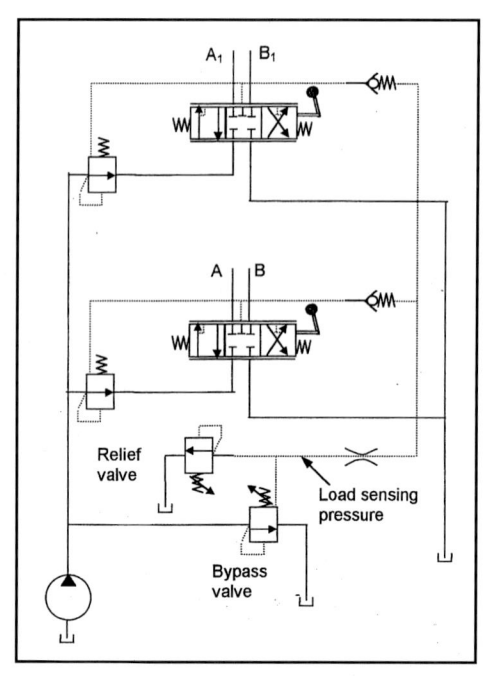

Figure 24. Pressure Compensated Bypass Valve with Load Sensing

The valve can be used to supply several actuators and, providing the pump flow is adequate, the flow to any selected service will not vary with changes in the required outlet pressure, i.e. it is not load sensitive. When the load sensing pilot pressure exceeds the setting of the relief valve it will open and the resulting flow will depress the pilot pressure on the bypass valve causing it to open so that the system flow is reduced and, hence the pressure.

6. Variable displacement pump control

There is a range of controls available for variable displacement pumps that are provided by most manufacturers to suit various application requirements. In general the pump displacement mechanism is operated by a hydromechanical servo that usually has the possibility of accepting an electrical input. In some cases the displacement is sensed by an electric transducer for closed loop control, which would normally be referred to as an electrohydraulic control system.

Circuit Design

6.1 Load sensing

For load sensing control the valve in Figure 24 can be used where the pilot signal is supplied to a valve that operates the pump displacement mechanism. Pressure limiting is usually incorporated so as to limit the maximum pump pressure. In applications where it is required to limit the input torque and power to the pump in order to prevent stalling of the prime mover, the servo system will, for a constant pump speed, provide a constant output power.

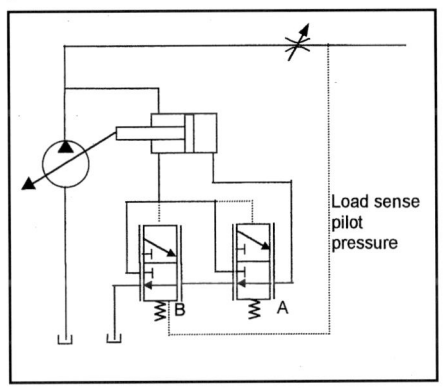

Figure 25. Variable Displacement Pump Pressure Limiting and Load Sensing Control

The circuit shown in Figure 25 controls the displacement to maintain the difference between the pump and the load sense pilot pressures. If the force from the pressure difference on valve B is higher than that set by the spring the pump displacement is reduced by the actuator until it has fallen to the correct value. Valve A is the pressure compensator that senses pump pressure, which, if it is too high, causes the pump displacement to be reduced in order to reduce the pressure.

6.2 Power control

For the control of the pump power Figure 26 shows a circuit that is used by Eaton for this purpose. This incorporates a pressure feedback from the hydraulic potentiometer that is varied by the pump displacement the force from which is opposed by that created by the pump outlet pressure as can be seen from Figure 26. If the pump pressure increases to too high a value its displacement is reduced by the action of the spool valve to a lower level that corresponds to that demanded for a constant power characteristic.

Figure 27 shows the operating envelope for a pump that has all three controls so that for a given flow demand, determined by the valve setting, the maximum available pressure will correspond to that at point A. As the flow demand is altered the maximum available pressure will be controlled to the constant power line. Thus point A traverses the maximum power line the maximum pressure being limited by the compensator setting.

Should the pump outlet pressure exceed this value the compensator will reduce the pump stroke and, hence, the flow. This control is able to reduce the pump displacement to zero. With a blocked outlet port the pump stroke reduces to a level when the output flow is just sufficient to make up the pump leakage at the set pressure.

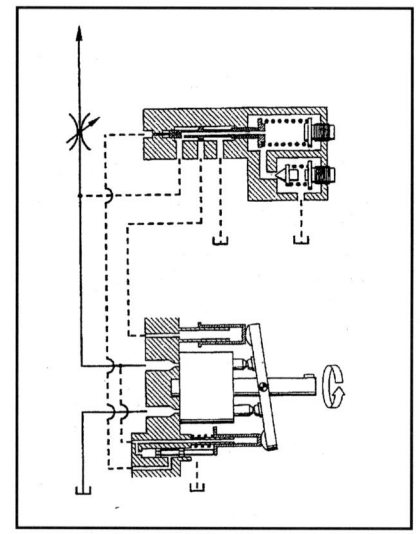

Figure 26. Constant Power Control (Eaton)

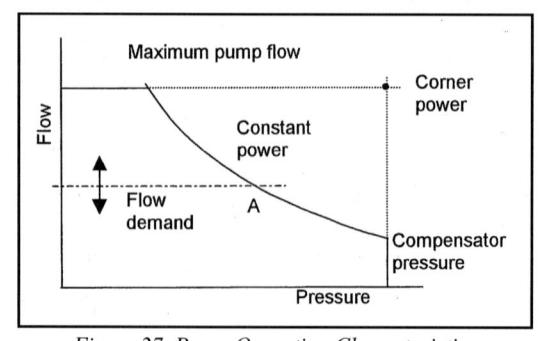

Figure 27. Pump Operating Characteristics

Circuit Design 101

6.3 Accumulator charging

Accumulators are frequently used in circuits to provide a supplementary flow in applications where the demanded flow varies in a cyclic manner. In this situation the accumulator can provide flow at a higher level than that available from the pump thus allowing a reduced capacity pump to be used.

This operation requires a circuit to:

- Increase the pump flow when the accumulator requires recharging
- Reduce the pump flow when there is sufficient fluid volume in the accumulator.

Figure 28 shows a circuit that is used to control the pump displacement to satisfy these requirements.

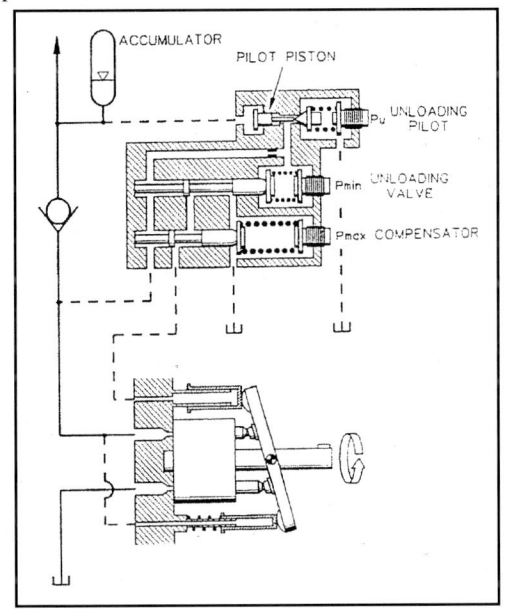

Figure 28. Accumulator charging circuit (Eaton)

After a period when the accumulator has discharged its volume and the unloading valve has closed, the pump displacement will increase and charge the accumulator. When the pump pressure reaches the value, P_U, set by the unloading pilot spring pre-load, the poppet will open. The associated flow through the restrictor will cause the unloading valve to move against its spring and pressurise the pump stroke piston to reduce the pump displacement.

The accumulator pressure will keep the unloading pilot poppet open because of the force on the pilot piston. The closed circuit check valve will maintain pressure in the accumulator circuit whilst the pump flow has reduced to zero. When the accumulator is required to discharge flow to the system, its outlet pressure will reduce until it has reached the level that causes the unloading pilot to close. The area of the pilot piston is greater than that of the poppet sensing area and, consequently, the level of pressure required to close the poppet can be typically 15% lower than that required by the pump pressure to open it.

7. Hydrostatic transmissions

7.1 Pump controlled systems (primary control)

Hydrostatic transmissions connect the actuator directly to the supply pump without using any valves for restrictive metering, the control of velocity being made by the displacement of the pump, and, in the case of motors, additionally by the motor displacement. The flow from the actuator is returned to the pump inlet thus avoiding the need for a large capacity boost pump.

The pressure level rises to that required to drive the actuator against the load. Consequently, the pump output flow can only be used to drive a single actuator or multiple actuators that are constrained to move at the same velocity (e.g. coupled motors and actuators attached rigidly to the same moving component)

Figure 29 shows the circuit for a hydrostatic transmission used to drive a hydraulic motor that includes:

- The provision of boost flow to make up for the external losses from the pump and motor. The check valves connect the boost input to the low pressure (unloaded) side of the loop. This applies for both the pump driving the motor and, for overrunning conditions, when the motor is driving the pump (e.g. winch lowering).
- Crossline relief valves to prevent excessive pressures. The flow is passed to the low pressure side in order to maintain the flow into the pump inlet.
- The extraction of fluid from the loop using a purge valve to provide increased cooling. This flow needs to be controlled as it has to be made up from the boost flow.

Circuit Design

- Variable motor displacement control for systems requiring a higher speed range at reduced torque.

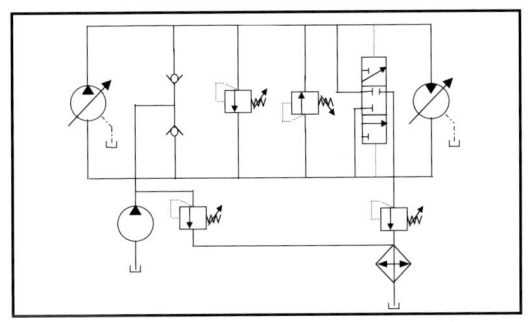

Figure 29. Rotary Hydrostatic Transmission Circuit

7.2 Motor brake circuit

The operation of spring loaded brakes is incorporated into the hydraulic system (e.g. winches, swing drives) by directing system pressure to the brake actuator as shown in Figure 30.

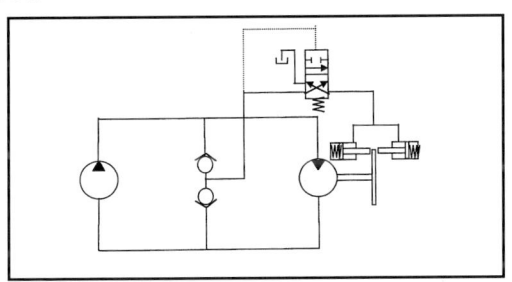

Figure 30. Motor brake circuit

The pressure required to release the brake actuators needs to be higher than the boost pressure. Often it is necessary to fit a reducing valve to limit the maximum pressure at the actuators if they have a lower rated pressure than the system.

7.3 Linear actuator transmissions

For linear actuator systems using equal area actuators the circuit is similar in principal to that for rotary systems as can be seen from Figure 31 which shows a basic circuit.

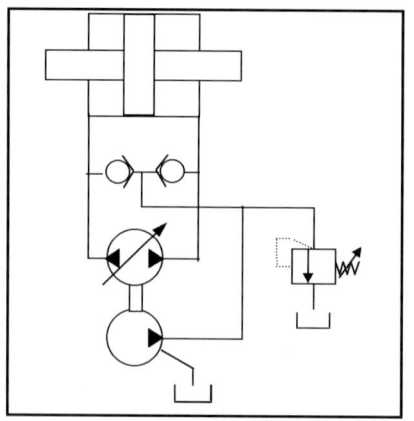

Figure 31. Linear Actuator Hydrostatic Transmission Circuit

7.4 Motor controlled systems (secondary control)

Secondary control systems operate at a constant supply pressure that can be provided by a pressure compensated pump the motor displacement being controlled so as to maintain constant speed in a closed loop system as shown in Figure 32.

The use of secondary control provides some advantages over the conventional hydrostatic system which include:

- The storage of energy in the accumulator from regenerative (e.g. overrunning) loads.
- Accuracy and dynamic performance.
- Use with multiple motors (ring main systems).

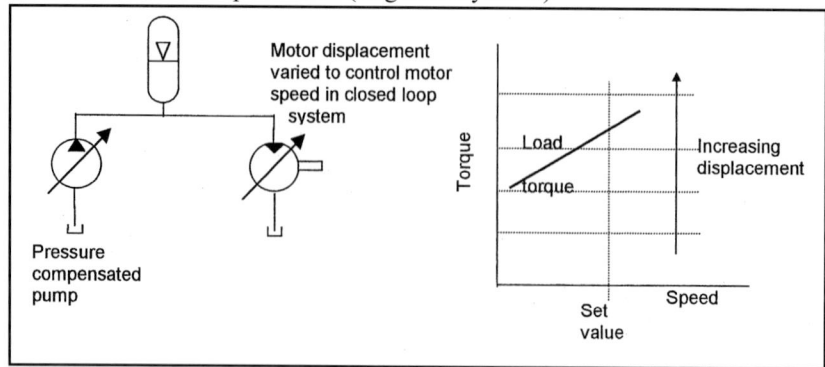

Figure 32. Secondary control system

Circuit Design 105

8. Pilot operated valve circuits

The pilot operation of valves has a wide range of application for the functional control in many machines and three control valve functions are described in this section.

8.1 Load control valves

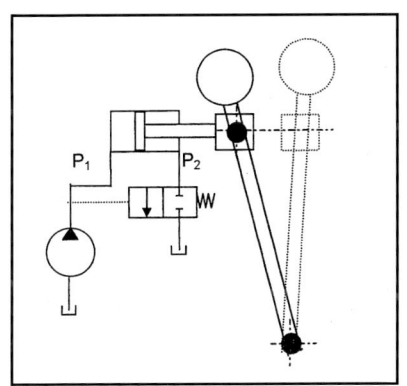

Figure 33. Load Control Circuit using Counterbalance Valves

In actuator systems when working with overrunning loads (e.g. crane booms, winches) it may be necessary to provide meter-out control in order to protect the system. Counterbalance valves are used for this purpose, as they require inlet pressure to the actuator to cause them to open. These act as closed loop systems and the dynamic behaviour during operation is complex and can often create oscillations of the load movement.

The circuit shown in Figure 33 is for operating a load that goes over centre and thus reversing the direction of the force acting on the actuator. As this force becomes negative it will initially cause the actuator velocity to increase which will reduce the inlet pressure P_1. When this pressure falls below the set value of the valve it will close so causing the actuator outlet pressure, P_2, to increase until it is at a sufficient level to resist the load force.

8.2 Pump unloading circuit

The double pump system in Figure 34 uses a pilot operated valve to connect one of the two pumps to tank when the load pressure exceeds the value set by the

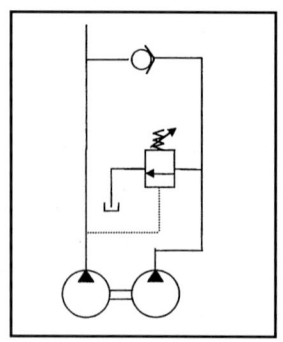

Figure 34. Double Pump System with Unloading Valve

spring. The outlet pressure from this pump will then be zero when the check valve will close and prevent flow from the high-pressure pump returning to the tank.

8.3 Sequence Control

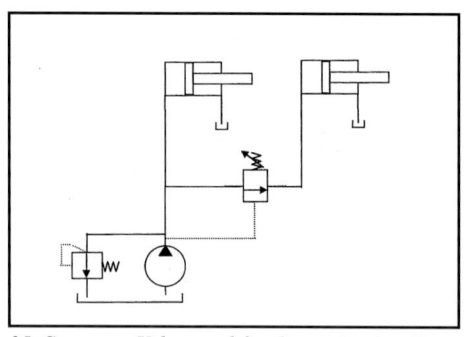

Figure 35. Sequence Valve used for Operation in a Press Circuit

The circuit in Figure 35 shows how pilot operated valves can be used to sequentially control the extensions of the two actuators. When one has reached the and of its stroke the increase in pump pressure that follows opens the sequential control valve thus allowing the second actuator to move.

9. Contamination control

Filters can be incorporated into hydraulic circuits in a number of ways, some of which are described in this section. Two basic points in the selection of a circuit depends on where the filter(s) are to be situated (high or low pressure) and the use of a filter bypass. The filter circuit shown in Figure 36 uses a high-pressure filter with a bypass check valve so that if the filter becomes blocked fluid can still be supplied to the system.

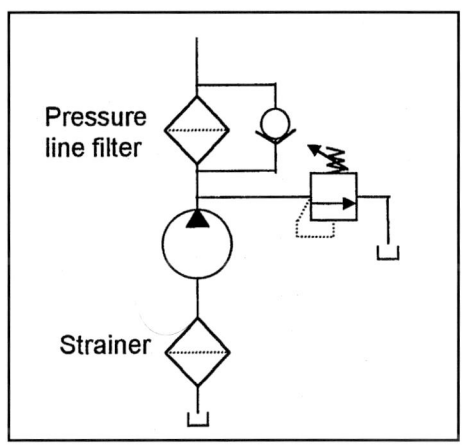

Figure 36. High pressure filter circuits

In situations where it is imperative that sensitive components are protected from contaminated fluid then the alternative approach is to not use a bypass. The system relief valve protects the pump from overpressures and, at the same time, the security of the filter housing. It is essential that in systems not employing a bypass the filter condition should be monitored carefully.

It is recommended that filters are not sited in areas of high vibration and, if possible, to put them in positions where the flow is constant. Both of these issues relate to the retention of contaminant in the filter that could otherwise become free to pass through the filter.

In order to maintain a constant flow through the filter the relief valve can be placed downstream of the filter as shown in Figure 37 (a).

Figure 37 (b) shows a low-pressure filter circuit that provides a cheaper alternative to high-pressure types. These filters are often of the spin-on type, which protect the reservoir and pump inlet from particles generated in the system but do not protect the system from particles generated in the pump. Spin-on filters can be sensitive to flow transients and pressure shocks.

The reservoir can be a major source of contamination and suction filters provide protection to the pump. However, because of the pressure loss in filters these can usually only provide filtration at levels of around 75 microns, the maximum allowable pressure loss being of the order of 0.2 bar. Higher pressure losses

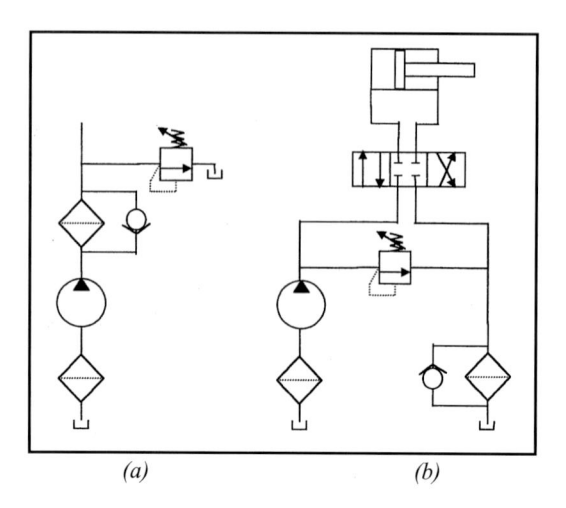

Figure 37. High and low pressure filter circuits

will cause aeration and cavitation of the fluid, and subsequent damage to the pump.

Off-line filtration can be used to circulate fluid from the reservoir on a continual basis. This system does not protect the components from contaminants created by a component failure but it does protect the hydraulic system on a long-term basis. The fluid is circulated by a pump that can also be used to top up, or fill, the reservoir with fluid that is pre-filtered by the off-line system.

The breather caps for the reservoir pass air in and out of the reservoir particularly in systems having a number of single-ended actuators. The breather should therefore be fitted with a filter that has the same level of filtration as the main system in order to prevent the ingress of large particulate contaminants.

CHAPTER EIGHT

FLOW PROCESSES
IN HYDRAULIC SYSTEMS

8. FLOW PROCESSES IN HYDRAULIC SYSTEMS

Summary

Pascal's Law for a fluid at rest states that the pressure at any point, neglecting head effects, is the same at any other point in the fluid. Generally speaking, in hydraulic systems, this law can be applied and the term 'hydrostatic' refers to this condition. However, when the fluid is moving this may not apply locally due to the effects of the fluid viscosity which create energy losses, that increase with the velocity. The fluid velocity also influences the pressure forces that act on components.

It is important that the system designer is aware of the background to the pressure/flow relationship in pipes, as there can be significant energy losses if their diameter is too small for the particular application. This relationship is based on the original theoretical and laboratory work carried out by Reynolds and an analysis of the pressure losses for piping is given in the Appendix.

For laminar flow the pressure loss is inversely proportional to the fluid viscosity, but this effect diminishes as the flow becomes more turbulent for Reynolds numbers >2000 when it tends to be proportional to the fluid velocity squared. The leakage of flow through small clearances such as are found around the pistons in pumps and motors can be similarly analysed to provide a basis for evaluating the variation of their volumetric efficiency with the fluid viscosity, pressure and rotational speed.

The pressure loss that occurs with flow in restrictions provides a principal method of controlling flow in hydraulic systems. For turbulent flow the flow through a spool valve is found to vary proportionally with its opening and with the square root of the pressure drop across the spool. This process is similarly important to the operation of pressure control valves such as relief valves and those used in many control circuits.

The level of the pressure force that acts on valves is strongly affected by the change in the momentum of the fluid as it passes through the valve. This process is difficult to analyse and is frequently analysed using Computational Fluid Dynamics (CFD) software. A simplified analytical method is discussed in this chapter which provides a basic understanding of the process that can be used to estimate these forces for some design purposes.

1. Introduction

This chapter is concerned with the analytical methods for the evaluation of the:

- Pressure losses in pipes and small clearances.
- Pressure/flow characteristics of restrictors.
- Fluid momentum forces on valves.

2. Fluid properties

Fluid density, ρ, is the mass per unit volume and has units of kg m^{-3}. The variation in the density of most hydraulic fluids with temperature is relatively small for the normal operating temperature usually encountered in hydraulic systems. For hydraulic oils the value can be taken as **870 kg/m^3**.

2.1 Fluid viscosity

The viscosity of the fluid can be referred to by:

i) **Dynamic viscosity, μ** for which the units are **Nsm^{-2}**.
The technical literature often uses other units. Typically **Poise (P)** where $1\ P = 0.1\ Ns\ m^{-2}$ and the **centiPoise (cP)** where $1\ cP = 10^{-3}\ Ns\ m^{-2}$.

ii) **Kinematic viscosity ν** This is more commonly quoted in technical literature and data sheets and is equal to the dynamic viscosity divided by the density. The units for kinematic viscosity are **m^2 s^{-1}** but it is usually quoted in **centiStoke (cSt)**

$$1\ cSt = 10^{-6}\ m^2\ s^{-1}.$$

The kinematic viscosity of a typical hydraulic oil varies with temperature and pressure as shown in Figure 1 and it can be seen that the viscosity varies considerably with changes in fluid operating temperature.

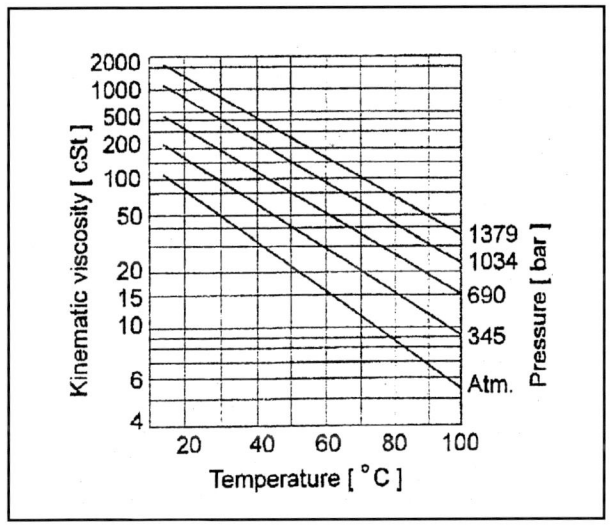

Figure 1. Oil viscosity variation with pressure and temperature

3. Flow in pipes

The effect that viscosity has on the flow through pipes results in a pressure loss that is determined by the Reynolds number, the length to diameter ratio of the pipe and the dynamic velocity head of the fluid. This pressure loss results in a loss of energy that is dissipated as heat in the fluid that may require cooling in some situations. Consequently, the loss of efficiency thus created has to be evaluated in order to establish the appropriate size of pipes, fittings and valves and the overall influence of the machine duty cycle on the fluid temperature.

The pressure loss in pump inlet, or suction, pipes is especially important as the flow is usually supplied from a reservoir at atmospheric pressure. This means, that in order to avoid cavitation and aeration of the fluid, the pressure loss must not exceed 0.2bar in most cases and in some instances it may be necessary to boost the pump inlet.

Air is normally absorbed in the fluid in the reservoir at atmospheric pressure and aeration occurs when the pressure falls below this value. When the fluid pressure falls to its vapour pressure, vapour bubbles begin to form. The problems caused by both of these processes, referred to as cavitation, are:

- Surface damage to the components
- Noise
- Damage and possible failure in pumps and motors
- Loss of control

The flow equation for the pressure loss, Δp, in a pipe of diameter d is given by:

$$\Delta p = 4f \frac{L}{d} \frac{1}{2} \rho u_m^2$$

This equation is derived in the Appendix (equation A5) for laminar flow where, f, is the friction factor which, for laminar flow, is given by:

$$f = \frac{16}{R_e} \text{ where } R_e \text{ is the Reynolds which} = \frac{\rho u_m d}{\mu} = \frac{u_m d}{\upsilon} \text{ and } u_m = \frac{4Q}{\pi d^2}$$

It is usual to employ the equation to determine the pressure loss in a pipe because, for a Reynolds number that is greater than 2000, the flow becomes turbulent and the velocity distribution is no longer parabolic as for laminar flow. For turbulent flow, the friction factor, f, has a different relationship with the Reynolds number and it is usual to obtain its value from the Moody chart shown in Figure 2.

Numerical example

Pipe flow $Q = 72 L/min = \dfrac{72}{6 \times 10^4} = 1.2 \times 10^{-3} m^3/s$

Pipe diameter $d = 25mm$; Area $A = \pi \times \dfrac{(25 \times 10^{-3})^2}{4} = 4.91 \times 10^{-4} m^2$

Flow velocity $u_m = ;\quad \dfrac{Q}{A} = \dfrac{1.2 \times 10^{-3}}{4.91 \times 10^{-4}} = 2.44 m/s;\ R_e = \dfrac{u_m d}{\upsilon}$

Consider values for the fluid viscosity of 70 and 20 cSt (i.e. 70 x 10⁻⁶ and 20 x 10⁻⁶ m²s⁻¹).

For the higher viscosity the value of R_e is 871 which is laminar when

$$f = \frac{16}{871} = 0.018$$

(this value can also be obtained from Figure 2).
For the lower viscosity, $R_e = 3050$ giving a value of $f = 0.011$ from Figure 2.
Considering a 10m-pipe length gives: $4\frac{L}{d}\frac{1}{2}pu_m^2 = 4.14 \times 10^6$

The pressure losses are 0.75 and 0.46 bar which would be considered low for flow in the high pressure side of a hydraulic system but would be unacceptable for the inlet supply to a pump.

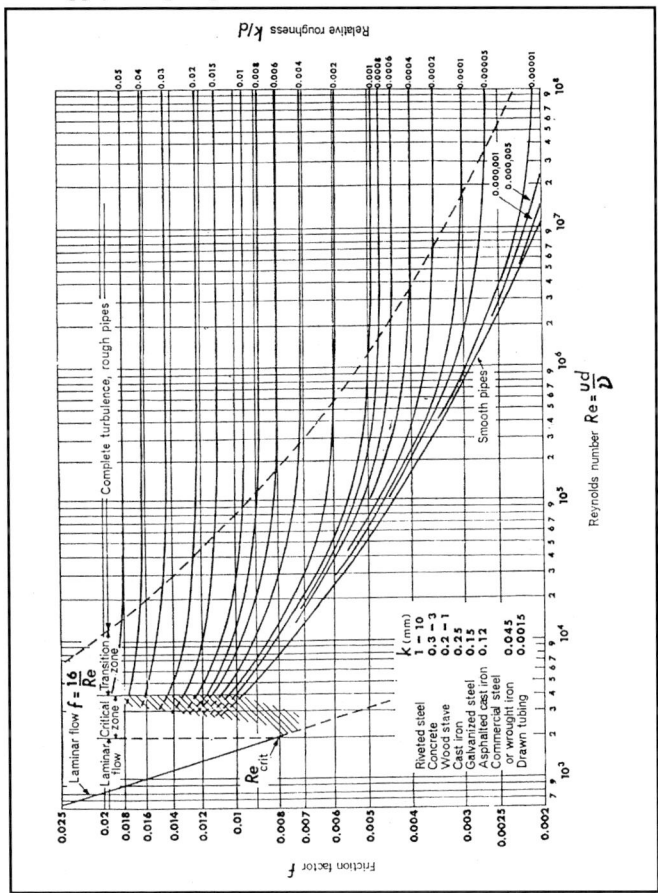

Figure 2. Moody Chart for determining the Friction Factor

4. Laminar flow in parallel leakage spaces

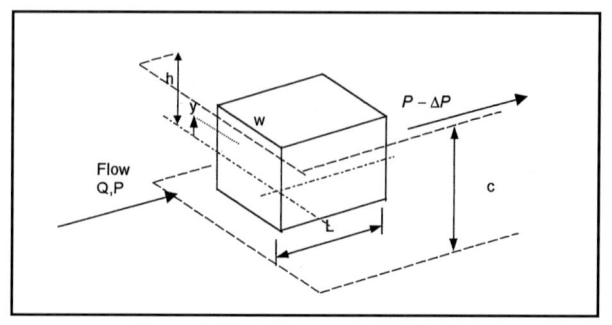

Figure 3. Flow between Parallel Plates

For fully developed steady laminar flow the velocity distribution with y across the gap between parallel plates is parabolic assuming that there is no transverse flow across the width of the flow path. In Figure 3, y is the vertical distance from the centre line in the element of fluid having a width w and length L. The analysis for this situation is similar to that used for the pipe in the Appendix where the total viscous force acting on the rectangular fluid element is given by:

$$\mu \frac{du}{dy} wL = -\Delta P w y$$

$$\therefore -\int_{u_1}^{0} du = \frac{\Delta P}{\mu L} \int_{0}^{h} dy$$

This gives for the maximum velocity, u_1, at the centre:

$$u_1 = \frac{\Delta P}{\mu L} \frac{h^2}{2} \qquad (1)$$

and

$$u = \frac{\Delta P}{2\mu L}(h^2 - y^2) \qquad (2)$$

From equation 2 the flow through the parallel path is given by:

$$Q = \int_{0}^{h} 2wu\,dy = \frac{w\Delta P}{\mu L}\left[h^2 y - \frac{y^3}{3}\right]_{0}^{h} = \frac{c^3 w}{12\mu L} \frac{\Delta P}{L} \qquad (3)$$

Flow Processes in Hydraulic Systems 117

Equation 3 is used to determine the leakage flow through small clearances between hydraulic components (e.g. spool valves, pump pistons and seals).

5. Orifice flow

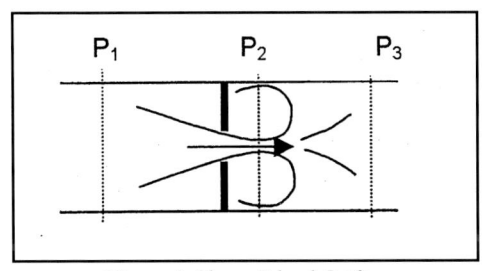

Figure 4. Sharp Edged Orifice

For flow through a sharp edged orifice the maximum velocity and lowest pressure is at the *vena contracta* that occurs downstream of the orifice itself the area of the *vena contracta* being smaller than that of the orifice. Assuming that the upstream velocity is very low in relation to that in the orifice the maximum velocity is obtained from the Bernoulli Energy Equation thus:

$$u = \sqrt{\frac{2(P_1 - P_2)}{\rho}} \qquad (4)$$

Flow, Q, through an orifice can be related to the area and the velocity. Generally:

$$Q = uA_0 \qquad (5)$$

where A_o = orifice area

Equation 5 gives the theoretical flow to which the discharge coefficient, C_d, is applied as in equation 6 to give:

$$Q = C_d A_o \sqrt{\frac{2(P_1 - P_2)}{\rho}} \qquad (6)$$

The pressure level, P_2, is not easily measured and it is usual to relate the flow through the orifice to the pressure drop from inlet to outlet and use a flow coefficient C_Q.

$$Q = C_q A_o \sqrt{\frac{2(P_1-P_3)}{\rho}}$$

Note that usually the orifice area is very much smaller than the upstream and downstream areas in which case:

$$C_q = C_d$$

For orifice Reynolds numbers higher than around 2000, the value of C_d for the sharp edge orifice tends to 0.62 but for valves having different geometrical configurations the value will depend on the particular shape. A typical variation of C_Q with Reynolds number is shown in Figure 5.

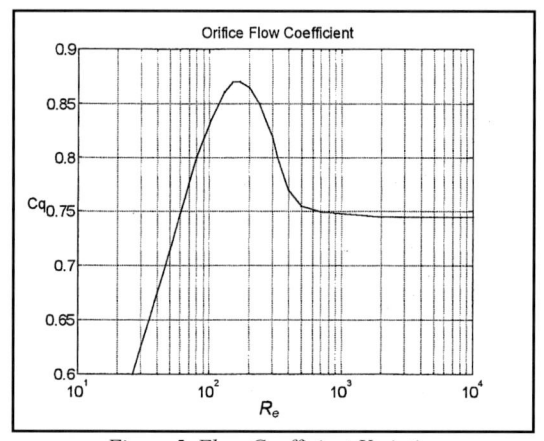

Figure 5. Flow Coefficient Variation

6. Valve force analysis

For flow passing through valve openings, the change in velocity is often high and the associated change in momentum has to be created by pressure forces in the fluid. The fluid generally enters the valve at a low velocity so that for a particle of fluid having a mass *m*, and an exit velocity *u*, the change in the momentum is given by *mu*.

The force required to create this change in momentum is given by:

$$f = m \frac{du}{dt}$$

Flow Processes in Hydraulic Systems 119

Now the mass is related to both the flow Q and the time period dt so that $m = \rho Q dt$. This gives for the rate of change of momentum:

$$f = \rho Q dt \frac{du}{dt} = \rho Q u$$

where u is the change in velocity.

6.1 Poppet valves

6.1.1 Momentum force

A force analysis for the poppet valve shown in Figure 6 needs to consider the pressure force on the valve face that will vary with the velocity of the fluid in the small valve opening.

With the valve in the closed position:

$$force = P_1 A_1$$

$$A_1 = \frac{\pi d^2}{4}$$

Figure 6. Single Stage Poppet Type Relief valve

For the valve open:

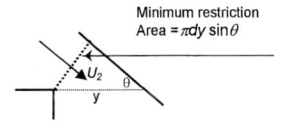

There is a reduction in the pressure as the velocity increases through the valve and increases the fluid momentum. The pressure force on the valve, which will depend on the pressure distribution on the valve face, is expressed by:

$$F = \int p dA$$

The upstream velocity will be low and hence the rate of increase in the axial momentum of the fluid through the valve is given by $\rho Q U_2 \cos\theta$. The pressure force on the fluid that is required to increase the fluid momentum is given by:

$$P_1 A_1 - \int p dA = \rho Q U_2 \cos\theta$$

Assuming that the downstream pressure P_2 is zero, the pressure force on the face of the valve must equal the spring force. Thus:

$$\int p dA = P_1 A_1 - \rho Q U_2 \cos\theta = C + ky \quad (7)$$

where C is the spring compression force when the valve is closed, k is the spring stiffness and $A_1 = \pi d^2/4$.

The term $\rho Q U_2 \cos\theta$ is referred to as the flow force (or Bernouilli force) which acts in the direction to close the valve and is additive to the force from the spring. As a consequence of this, in order to increase the flow through the valve, the upstream pressure force $P_1 A_1$ has to increase.

6.1.2 Valve flow

The restriction created by the valve reduces the pressure of the fluid. Assuming that P_2 is zero the valve flow equation gives:

$$Q = C_Q \pi d y \sin\theta \sqrt{\frac{2}{\rho} P_1}$$

$$\therefore y = \frac{Q}{C_Q \pi d \sin\theta \sqrt{\frac{2}{\rho} P_1}} \quad (8)$$

Also

$$U_2 = \sqrt{\frac{2}{\rho} P_1} \quad (9)$$

6.1.3 Valve pressure/flow characteristics

Combining equations (7), (8) and (9) gives:

$$P_1 = \frac{C}{A_1} + K_1 \frac{Q}{\sqrt{P_1}} + K_2 Q \sqrt{P_1} \quad (10)$$

Where:
$$K_1 = \frac{k}{A_1 \pi d C_Q \sin\theta \sqrt{\frac{2}{\rho}}}$$

and
$$K_2 = \frac{\cos\theta}{A_1}\sqrt{2\rho}$$

$$\therefore Q = \frac{P_1 - C/A_1}{\frac{K_1}{\sqrt{P_1}} + K_2\sqrt{P_1}} \tag{11}$$

It is seen from equation 11 that the force coefficient K_2 (due to the flow force) has an additive effect to the force coefficient K_1 that arises from the spring stiffness. Increasing the values of these coefficients reduces the level of flow that the valve will pass for a given upstream pressure P_1. The valve flow to pressure characteristics obtained from equation 11 are shown typically in Figure 7 where the zero flow intercept is referred to as the 'cracking pressure'.

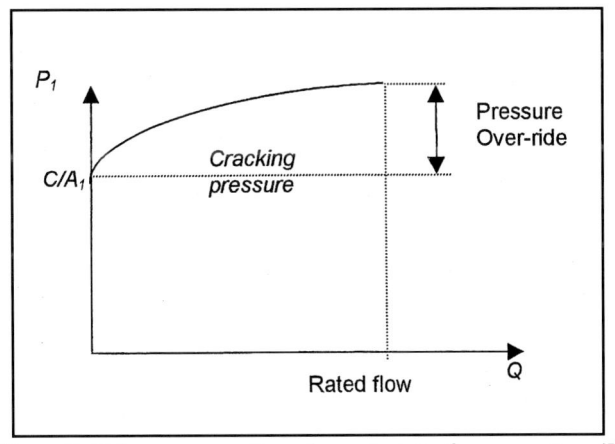

Figure 7. Valve pressure flow characteristic from equation (5)

6.2 Spool valves

In spool valves, there is a peripheral flow of fluid through the annulus formed by the valve land and the port.

```
         ┌─────────────────────────────┐
         │        High    │ /φ         │
         │      Pressure  │/           │
         │        │Pₛ│   /             │
  ┌──────┤                             ├──────┐
  │      │                             │      │
  │      │                             │      │
  └──────┤             ↓               ├──────┘
         │             →               │
         └─────────────────────────────┘
```

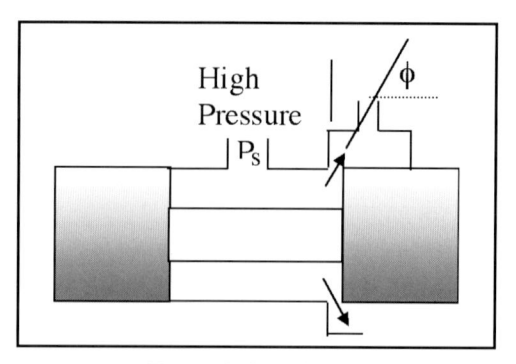

Figure 8. Spool valve

The flow from the high pressure inlet to the spool valve in Figure 8 creates a high velocity in the outlet metering annulus so that there is a resulting force on the spool tending to close the valve. This arises from the difference between the pressure force on the spool land on the left side, which is due to the inlet pressure, P_S, and that on the right hand spool land, which is less because the pressure is reducing radially outwards due to the increasing fluid velocity. This pressure force is equal to the axial component of the momentum change so for a velocity U, the momentum force is given by:

$$\rho Q U \cos\phi$$

The flow through the restriction is given by:

$$Q = A_0 U = C_Q A_0 \sqrt{\frac{2P_S}{\rho}}$$

where A_0 is the valve opening area = $\pi d x$ and x is the valve displacement. Thus the momentum force = $2\pi d C_Q x P_s \cos\phi$. The angle ϕ is usually taken as 69^0 (Von Mises) but this can vary as a function of the valve clearance and other dimensional parameters. (Ref: Control of Fluid Power Analysis and Design, 2nd (Revised) Edition, D. McCloy and H. R. Martin; Elis & Horwood Ltd; ISBN: 0-85312-135-4 [Out of Print]).

Appendix

This appendix provides the background to the equations that are used for determining the relationship between flow and pressure loss in pipes.

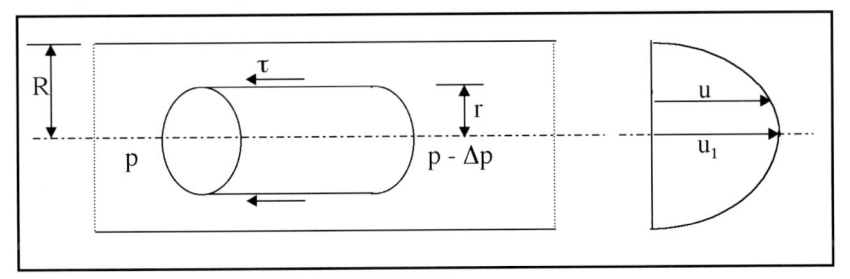

Figure A1. Flow through a Pipe

For fully developed, steady laminar flow the theoretical velocity distribution is parabolic which is verified by experiment. Consider a cylindrical element of fluid of radius r and length L. The total viscous force acting on the cylindrical element is given by:

$$2\pi r L \tau = -2\pi r L \mu \frac{du}{dr}$$

μ = fluid dynamic viscosity
τ = shear stress in the fluid

For steady motion of the fluid (i.e. no change in the velocity):

$$\pi r^2 \Delta p = -2\pi r L \mu \frac{du}{dr}$$

and:

$$-\int_{u_1}^{0} du = \frac{\Delta p}{2\mu L} \int_{0}^{R} r\, dr$$

The velocity is zero when $r = R$ and is a maximum when $r = 0$. The solution of the integrals gives:

$$u_1 = \frac{\Delta p}{4\mu L} R^2 \qquad (A1)$$

For any radius r in the range, $0 \leq r \leq R$ where $u_1 \geq u \geq 0$ the velocity distribution is given by:

$$u = \frac{\Delta p}{4\mu L}(R^2 - r^2) \qquad (A2)$$

The total volume flow Q is given by:

$$Q = \int_0^R 2\pi r u \, dr = \frac{\Delta p}{8\mu L}\pi R^4$$

The mean velocity, u_m, is given by:

$$u_m = \frac{Q}{\pi R^2} = \frac{\Delta p}{8\mu L}R^2 = \frac{u_1}{2} \qquad (A3)$$

Therefore:

$$\Delta p = \frac{8\mu u_m L}{R^2} = \frac{32\mu u_m L}{d^2} \qquad (A4)$$

It is normal to relate this to the Reynolds number Re:

$$R_e = \frac{\rho u_m d}{\mu}$$

By rearranging equation A4 we get:

$$\Delta p = \frac{32\mu u_m}{d} \times \frac{L}{d} \times \frac{\rho u_m}{\rho u_m} = 4(\frac{16\mu}{\rho u_m d}) \times \frac{L}{d} \times \frac{1}{2}\rho u_m^2 = 4f\frac{L}{d}\frac{1}{2}\rho u_m^2 \quad (A5)$$

The parameter, f, is referred to as the friction factor and, for laminar steady flow it is given by $\frac{16}{R_e}$.

The term $\frac{1}{2}\rho u_m^2$ is referred to as the dynamic, or velocity head.

Reference

Fluid Mechanics and Heat Transfer, J M Kay, Cambridge University Press, 1957

CHAPTER NINE

OPERATING EFFICIENCIES OF PUMPS AND MOTORS

9. OPERATING EFFICIENCIES OF PUMPS AND MOTORS

Summary

The performance of pumps and motors is subject to the effect of torque and leakage losses that determine their overall efficiency. Manufacturers' technical literature will usually show the efficiency variation over the range of operating conditions for a particular value of the fluid viscosity. For the use of the unit with a fluid having a different viscosity from that quoted requires a knowledge of the way in which the mechanical and volumetric losses arise. The processes involved can be modelled using the analytical methods described in this chapter which also provides a background for their implementation in computer data base and simulation systems.

1. Introduction

The leakage losses in pumps and motors are derived analytically using the method developed by Wilson[1] that is based on the laminar flow equations in chapter 8 which produces a simple model for the volumetric efficiency. The mechanical losses are based on assuming that viscous forces between moving components are proportional to the viscosity and the relative velocity and that there is a coulomb friction force that is proportional to the pressure.

2. Mechanical and volumetric efficiency

The power loss mechanism is the same for both pumps and motors. However, the efficiency of pumps and motors are slightly different because of the way the losses affect the power flow in the units. In pumps, to achieve a given output power in the fluid, the power losses have to be added to the theoretical power at the input shaft so that as the overall efficiency reduces, the input power has to be increased. In motors the power losses detract from the input fluid power and the result is that the same level of losses give a different efficiency.

Denoting the volumetric, or leakage, flow loss by Q_s and the mechanical or frictional torque loss by T_m gives the efficiencies as:

For a pump:

$$\eta_v = \frac{Q_t - Q_s}{Q_t} = 1 - \frac{Q_s}{Q_t} \tag{1}$$

$$\eta_m = \frac{T_t}{T_t + T_m} = \frac{1}{1 + T_m/T_t} \tag{2}$$

And for motors:

$$\eta_v = \frac{Q_t}{Q_t + Q_s} = \frac{1}{1 + Q_s/Q_t}$$

$$\eta_m = \frac{T_t - T_m}{T_t} = 1 - \frac{T_m}{T_t}$$

In these equations the ideal flow is denoted by Q_t, and the ideal torque by T_t. For either unit, the overall efficiency is given by:

$$\eta_o = \eta_v \eta_m$$

These losses can be determined from test results but in order to establish their variation over the range of operating conditions requires a theoretical approach such as that described in the next section.

3. Analysis of the losses

3.1 Theoretical performance

For the ideal pump, or motor, with no losses, the ideal or theoretical flow, Q_t depends only on the pump geometric capacity, D and its rotational speed, ω, hence:

$$Q_t = D\omega$$

Operating Efficiencies of Pumps and Motors   129

Equating the mechanical power with the fluid power gives:

$$T_t \omega = PQ_t$$

Hence the ideal or theoretical torque T_t depends only on the displacement and the differential pressure across the unit.
Thus:

$$T_t = DP$$

At the system design stage the theoretical equations can be used but when units have been selected actual efficiency values can be obtained from the technical literature.

Most mathematical models for pump and motor steady state performance are based on the work of W.E. Wilson.

3.2 Volumetric flow loss

The volumetric flow loss arises from leakage through the various clearance spaces between the moving components in the unit. The degree of these losses will vary between the different types of units but in this analysis the overall leakage is assumed to be dependent on the fluid pressure and viscosity as obtained in section 4 of Chapter 8.

Thus the laminar flow equation for flow through a slot gives:

$$Q_s = \frac{h^3}{12\mu} w \frac{dp}{dx}$$

Here μ is the fluid dynamic viscosity. If the clearance, h, is assumed to be constant, for a given unit this equation can be reduced to:

$$Q_S = C_1 \frac{P}{\mu}$$

And the ratio,

$$\frac{Q_S}{Q_t} = \frac{C_1}{D}\left(\frac{P}{\mu\omega}\right) \qquad (3)$$

Thus it is seen that the flow ratio, $\frac{Q_l}{Q_s}$, is a function of a non-dimensional parameter, $\frac{P}{\mu\omega}$, that will tend to zero for low pressures, high speeds and fluids with a high viscosity i.e. the leakage will be zero. Whilst the value of C_1 can be obtained from test data these results usually tend to have significant levels of scatter because:

- The clearances are not likely to be constant during the rotating cycle e.g. a piston can be eccentrically disposed and/or tilted in a cylinder bore (for contact conditions the leakage increases 150% over that when concentric).
- The component sliding component velocities are not constant during the rotating cycle.
- The fluid viscosity will vary through the leakage path due to frictional heating.
- There may be a variation in the leakage with the rotational speed.

In many unit types the leakage loss has been shown to vary as $P^{1.5}$. In the above equations, P is usually taken as the difference between the inlet and outlet pressures for a pump or motor. However, in piston units much of the leakage is to the case drain, which may not be at the same pressure as the pump inlet or motor outlet.

4. Mechanical loss

Mechanical losses occur as a result of viscous and coulomb friction. Viscous friction arises from fluid shear in the clearance spaces that result from the relative speed of the various moving components and is referred to as speed dependent friction. Viscous friction torque is expressed by:

$$T_v = C_2 \mu \omega$$

And the torque ratio,
$$\frac{T_v}{T_t} = \frac{C_2}{D}\left[\frac{\mu\omega}{P}\right] \tag{4}$$

This torque ratio is a function of a non-dimensional parameter, $\frac{\mu\omega}{P}$, that is the inverse of that for the flow ratio. The value of the coefficient C_2 can be obtained from tests but the results will also be dependent on the same parameters identified as being likely to have an influence on C_1.

Coulomb friction torque is proportional to load and independent of speed:

$$T_f = C_3 P \tag{5}$$

Operating Efficiencies of Pumps and Motors 131

and the torque ratio:

$$\frac{T_f}{T_t} = \frac{C_3}{D} \qquad (6)$$

The value of the coefficient C_3 depends on the frictional conditions between the various sliding components. In some gear units the coulomb friction torque loss can vary with $P^{1.5}$ and in all types of unit the relative frictional loss increases at low speeds.

The total mechanical loss,

$$T_m = T_V + T_f \qquad (7)$$

5. Unit efficiency

5.1 Volumetric efficiency

For pumps the variation of the volumetric efficiency with the non-dimensional parameter $\frac{P}{\mu\omega}$ can be obtained by substituting equation (3) into (1) which gives:

$$\eta_v = \frac{Q_t - Q_s}{Q_t} = 1 - C_s \left[\frac{P}{\mu\omega} \right] \qquad (8)$$

Where $C_s = \frac{C_1}{D}$

Here it is seen that as the pressure is increased and the speed and fluid viscosity are reduced, the volumetric efficiency is reduced. Pumps are tested at given operating conditions and equation (8) can be used to estimate the volumetric efficiency at different operating conditions.

5.2 Mechanical efficiency

The mechanical efficiency can be obtained similarly by substituting equations (4) and (6) into equation (2) to give:

$$\eta_m = \frac{T_t}{T_t + T_V + T_f} = \frac{1}{1 + C_f + C_v \left[\frac{\mu\omega}{P} \right]} \qquad (9)$$

Where $C_f = \dfrac{C_3}{D}$ and $C_v = \dfrac{C_2}{D}$

The mechanical efficiency from equation (9) is seen to vary in the opposite manner with pressure speed and viscosity to the volumetric efficiency.

5.3 Overall efficiency

The overall efficiency, η_0, which is the product of the volumetric and mechanical efficiencies will vary with the non-dimensional parameter as shown Figure 1.

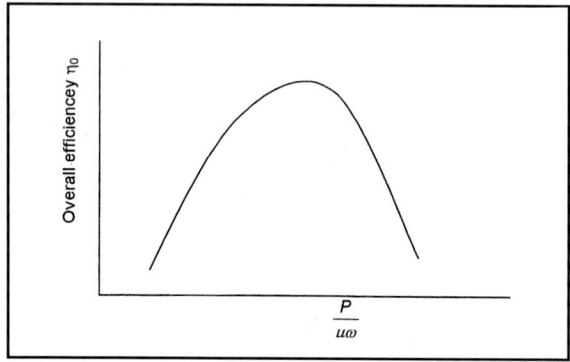

Figure 1. Variation of Overall Efficiency

Figure 1 shows that pumps (and motors) have an optimum operating condition which is a combination of the pressure, speed and fluid viscosity. This condition can be theoretically determined by considering the total power loss in the unit, which is given by:

$$\dfrac{Power\ loss}{Ideal\ power} = C_s \left[\dfrac{\Delta P}{\mu \omega}\right] + C_f + C_v \left[\dfrac{\mu \omega}{\Delta P}\right]$$

Differentiating with respect to [$\dfrac{\mu \omega}{\Delta P}$] and equating the expression to zero gives the minimum power loss when:

$$-C_s \left[\dfrac{\mu \omega}{\Delta P}\right]^{-2} + C_v = 0 \quad or \quad \dfrac{\mu \omega}{\Delta P} = \sqrt{\dfrac{C_s}{C_v}}$$

At this condition, the slip and viscous friction losses are equal. This suggests that for maximum overall efficiency, the volumetric losses and mechanical losses should be of similar magnitude; alternatively that η_v and η_m should be similar.

This method has wide applications in optimising designs and operating conditions.

Reference

1. Rotary pump theory, Wilson W.E., Trans ASME 1946, 68, 371.

CHAPTER TEN

CONTROL SYSTEM DESIGN

10. CONTROL SYSTEM DESIGN

Summary

Hydraulic power is used extensively in closed loop control systems because the high force to mass ratio provides a relatively high natural frequency and fast response. Electrohydraulic control allows flexibility in terms of the range of parameters that can be controlled (e.g. position, velocity, pressure and force) and also has the capability to incorporate elements that assist in improving the steady state and dynamic accuracy and the stability margin of a system.

This chapter develops the equations that describe the dynamic performance of actuator control systems, which are used to determine their frequency response and the stability margins so that appropriate components can be selected. Because of non-linear characteristics the analysis is based on the small perturbation technique which enables the dynamic performance to be predicted. However, predicting the steady state performance has to take account of the various non-linearities and some compensation methods for improving this and the dynamic performance are discussed.

The system can be approximated to a first order response by assuming the load has a negligible mass so, having selected an appropriate actuator and the valve size to provide the necessary flow and actuator velocity, the time constant can be determined. The effect of load mass introduces a second order mass/spring/damper into the open loop transfer function having a natural frequency that is a function of the fluid bulk modulus, the inertial mass, the actuator area and the compressed volume. This frequency enables an approximate value of the system gain to be determined using a simple design procedure that assumes a value for the damping ratio.

The transfer function obtained from the linearised equations provides a good estimate of the system performance for small changes but for high accuracy and/or fast response compensation circuits have to be incorporated into the electronic control. These can be made from either electronic components or can be incorporated into computer software where appropriate.

1. Introduction

Hydraulic valves are extensively employed in the closed loop position of actuators so as to obtain good steady state accuracy and speed of response. The open loop velocity control of linear actuators was discussed in Chapter 7 where the flow through the valve, for constant supply pressure, was shown to depend upon the valve opening and the pressure drop across the opening. Thus, under specified pressure conditions, the valve can be used as a flow source when connected to an actuator the velocity of which will be proportional to the valve opening.

The dynamic performance of a valve actuator system is developed in which it is assumed that the mass of the moving components and the load force are both zero. The effect of these assumptions is to eliminate the influence of the fluid compressibility so that the system response is first order.

The inclusion of the mass into the analysis creates a third order system because of the effect of the fluid compressibility and, as a consequence of this, the value of the system gain has to be determined in order to obtain satisfactory stability margins for the system.

2. Simple valve actuator control

2.1 Open loop system

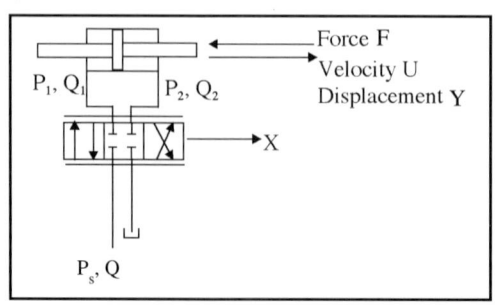

Figure 1. Valve Actuator Circuit

Control System Design 139

The valve actuator circuit in Figure 1 has an equal area, or double ended, actuator that is frequently used for closed loop control systems because of its symmetry with regard to the hydraulic force and velocity in both directions of movement. The force characteristics for the system can be obtained from the analysis in Chapter 7, equations 3 to 6, which with the area ratio $\alpha = 1$ gives:

<table>
<tr><th>Extend</th><th>Retract</th></tr>
</table>

$$\frac{P_1}{P_s} = \frac{(1+R\alpha^3)}{(1+\alpha^3)} \qquad P_1 = \frac{P_s}{2}(1+R) = \frac{P_s}{2} + \frac{F}{2A} \qquad \frac{P_1}{P_s} = \frac{\alpha^2(1+\alpha R)}{(1+\alpha^3)} \qquad P_1 = \frac{P_s}{2} + \frac{F}{2A}$$

$$P_1 = \frac{P_s}{2} \text{ for } F = 0 \qquad\qquad P_1 = \frac{P_s}{2} \text{ for } F = 0$$

$$\frac{P_2}{P_s} = \frac{\alpha(1-R)}{(1+\alpha^3)} \qquad P_2 = \frac{P_s}{2}(1-R) = \frac{P_s}{2} - \frac{F}{2A} \qquad \frac{P_2}{P_s} = \frac{\alpha^3 - \alpha R}{1+\alpha^3} \qquad P_2 = \frac{P_s}{2} - \frac{F}{2A}$$

$$P_2 = \frac{P_s}{2} \text{ for } F = 0 \qquad\qquad P_2 = \frac{P_s}{2} \text{ for } F = 0$$

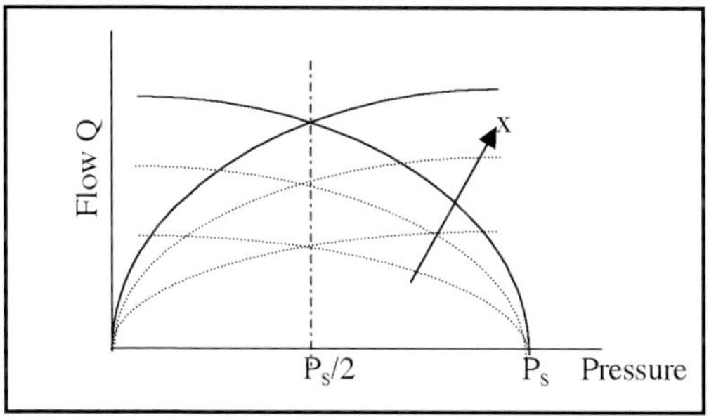

Figure 2. Valve Characteristics

The force F is positive for the direction shown in Figure 1. For the equal area actuator, the valve flows are the same on each side of the valve and the pressure differences, $(P_s - P_1)$ and P_2, respectively, will therefore be the same.

For zero force, $P_1 = P_2 = P_s/2$ that, as can be seen from Figure 2, is the pressure at which the flow characteristics intersect.

The flow through the valve is given by:

$$Q_{1,2} = Q = C_Q \pi dX \sqrt{\frac{2}{\rho} \frac{P_s}{2}}$$

For the simple system assume that the moving components have negligible inertia so that, as a consequence, the actuator pressures will remain constant during transient changes caused by displacement of the valve.
Thus:

$$Q = K_Q X \tag{1}$$

where: $K_Q = C_Q \pi d \sqrt{\dfrac{P_s}{\rho}} \ m^2/s$

And, for an actuator area A the actuator velocity, U is:

$$U = \frac{Q}{A} = \frac{K_Q X}{A} \tag{2}$$

As $U = \dfrac{dY}{dt}$, then $Y = \dfrac{K_Q}{A} \int X dt$ (3)

The actuator displacement is the integral of the valve opening and, as a consequence, it is referred to as an *integrator*. A step opening of the valve will, therefore, cause the actuator to move at a constant velocity, stopping when the valve is closed. This describes the *open-loop* performance of the system.

The time response of the actuator, or integrator, to a step change in valve position is shown in Figure 3.

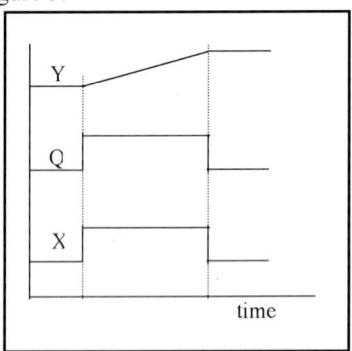

Figure 3. Open loop time response

2.2 Closed loop system

Closed loop systems operate by comparing the output position with an input demand signal such systems being referred to as feed back control. In

Control System Design 141

electrohydraulically operated valve actuator systems, the input signal is a voltage and the output position is fed back as a voltage signal from a position transducer. The comparison of the two signals produces a voltage difference referred to as the error signal that is amplified as a current signal and supplied to the electrohydraulic valve. In this analysis the relationship between the input and output is obtained from the following equations.

The valve position is given by:

$$X = K_A K_V (V_i - K_T Y) \qquad (4)$$

K_T = position transducer gain, V/m
K_V = valve gain, m/A
K_A = amplifier gain, A/V
V_I = input signal, V

Equations 2, 3 and 4 give:

$$\frac{dY}{dt} = \frac{K_Q}{A}(K_A K_V (V_i - K_T Y))$$

It is usual to express $\frac{d}{dt}$ by the Laplace operator 's' allowing this equation to be treated algebraically:

Thus: $\quad (K_T + \frac{A}{K_Q K_A K_V} s)Y = V_i$

This gives: $\quad \dfrac{Y}{V_i} = \dfrac{1/K_T}{(1 + Ts)} \qquad (5)$

This is a first order differential equation where the time constant $T = \dfrac{A}{K_A K_Q K_V K_T}$ which has the units of time (s).

The system can be represented as a block diagram:

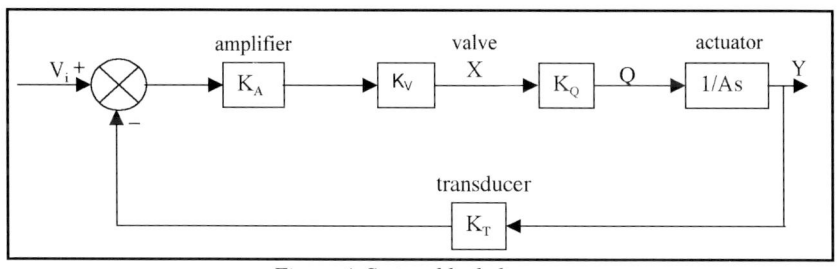

Figure 4. System block diagram

This diagram can be simplified.

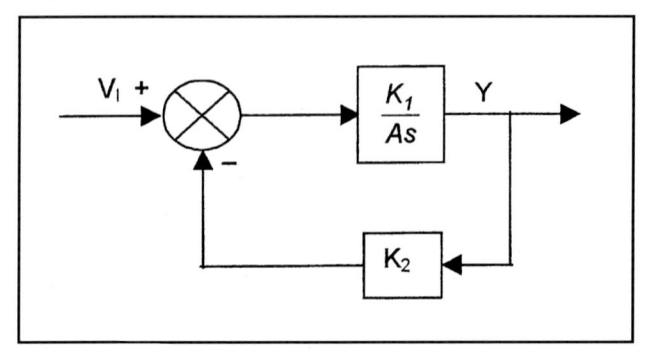

Figure 5. Simplified block diagram

Note that electrohydraulic manufacturer's literature will usually specify the valve performance in terms of flow for a given input current at a specific valve pressure drop, which will enable the gain product $K_V K_Q$ to be obtained.

The transfer function for the relationship between y and q_i can be obtained from Figure 5 which is given by:

$$\frac{y}{V_i} = \frac{K_1/As}{1 + K_1 K_2 / As} = \frac{1/K_2}{1 + Ts} \quad \text{where } T = \frac{A}{K_1 K_2} \quad (6)$$

Substituting for K_1 and K_2 in equation 6 gives the same result as in equation 5.

2.3 System response

The time response from equation 5 can be obtained for a range of input signals by referring to a dictionary of standard inverse Laplace transforms. The response to a step input gives the solution:

$$\Delta F = \Delta P_1 A_1 - \Delta P_2 A_2 \quad (7)$$

This exponential variation of Y with time reaches 63.2% of the final value in one time constant as shown in Figure 6.

Control System Design   143

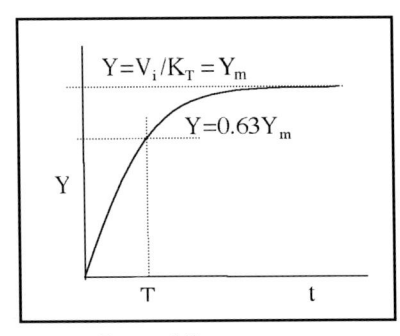

Figure 6 Step response

3. Fluid compressibility

3.1 Bulk modulus

The compressibility of fluids is expressed by the bulk modulus, β, which varies with pressure and, as a consequence, either average or local values have to be used. However, the bulk modulus is also considerably affected by absorbed air in the fluid and also by the use of hoses. As a consequence the value used in the analysis has to be considered carefully.

Bulk modulus is defined as

$$\beta = -\frac{P}{\Delta V / V}$$

where the term $\frac{\Delta V}{V}$ is the volumetric strain in the fluid. For hydraulic oil the value of β is usually taken as 1.8×10^9 N/m² but this can reduce by 30 to 50% in some circumstances.

3.2 Hydraulic stiffness

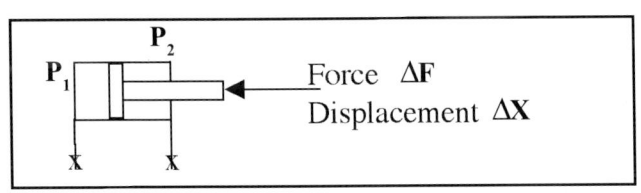

Figure 7. Fluid compressibility

With both ports blocked as in Figure 7 and with an initial pressure level that will prevent negative pressures from arising we get:

Pressure changes from the actuator force $\Delta F = \Delta P_1 A_1 - \Delta P_2 A_2$ (7)

From the bulk modulus $\Delta V_1 = -V_1 \dfrac{\Delta P_1}{\beta}$ and $\Delta V_2 = -V_2 \dfrac{\Delta P_2}{\beta}$ (8)

Also $\Delta V_1 = -A_1 \Delta X$ and $\Delta V_2 = A_2 \Delta X$ (9)

Equations 7, 8 and 9 give: $\dfrac{\Delta F}{\Delta X} = \beta \left(\dfrac{A_1^2}{V_1} + \dfrac{A_2^2}{V_2} \right)$ (10)

For an equal area actuator with $A_1 = A_2 = A$ and with the piston in the central position with $V_1 = V_2 = V$ equation 10 gives for the hydraulic stiffness:

$$\dfrac{\Delta F}{\Delta X} = 2 \dfrac{\beta A^2}{V} = K_H \quad (11)$$

The bulk modulus has two significant effects on the dynamic performance of closed loop hydraulic systems in that:

- The hydraulic stiffness gives rise to a hydraulic natural frequency:

$$\omega_n = \sqrt{\dfrac{\text{stiffness}}{\text{mass}}} = \sqrt{\dfrac{K_H}{m}} \text{ rad / s}$$

- The compressibility creates transient delays in volumes where

$$\dfrac{dP}{dt} = \dfrac{\beta}{V} \dfrac{dV}{dt} = \dfrac{\beta}{V} Q_C$$

(Q_C is the difference between the inlet and outlet flow to the volume V).

4. Valve actuator dynamic response including compressibility effects

4.1 Valve flow

The simple system model assumed that the inertial mass of the moving components was zero so that the actuator pressures remained constant during transients.

Control System Design 145

In the presence of inertial load mass, the pressures will have to change from the steady state values in order to change the velocity of the actuator and these changes will need to take account of the fluid compressibility in relation to the system flows.

The analysis in 2.1 can be applied to systems where the force is not zero by determining the steady state actuator pressures. For the equal area actuator during extension as in Figure 1 these are:

$$P_1 = \frac{P_s}{2} + \frac{F}{2A} = \frac{P_s}{2} + \frac{P_m}{2} \qquad (12)$$

and

$$P_2 = \frac{P_s}{2} - \frac{F}{2A} = \frac{P_s}{2} - \frac{P_m}{2} \qquad (13)$$

The pressure P_m is the load pressure and it is seen that it is divided equally in its effect on P_1 and P_2 which will affect the flows also equally as shown in Figure 8.

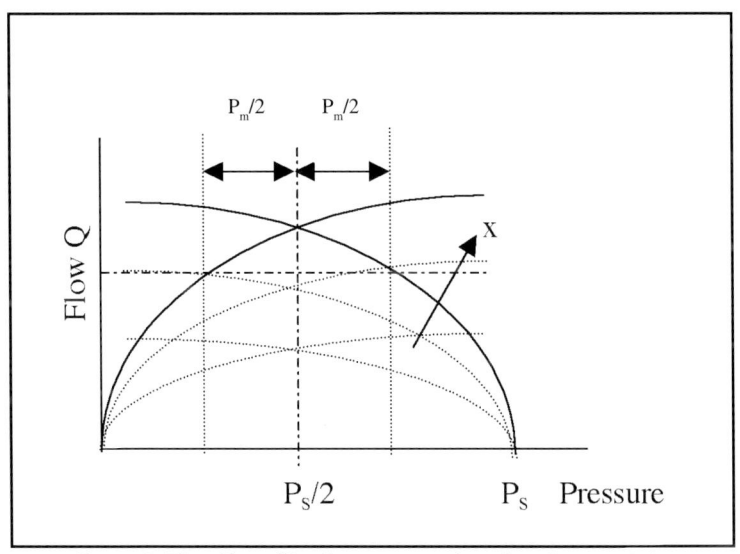

Figure 8. The effect of load pressure on the valve characteristics

The load force will affect the flow coefficient K_Q in equation 1 because the valve pressure differences become:

$$P_s - P_1 = \frac{1}{2}(P_s - P_m) = P_2 \qquad (14)$$

Equation 1 can be restated as:

$$Q = C_Q \pi dX \sqrt{\left(\frac{P_S - P_m}{\rho}\right)} = K_Q X \sqrt{\left(1 - \frac{P_m}{P_S}\right)} \quad (15)$$

For $P_m = 0$ (i.e. zero force) equation 15 reverts to equation 1. The load pressure ratio P_m/P_S varies in the range $+1$ to -1 (pushing to pulling) and from equation 15 it is seen that the maximum flow is $K_Q X$.

For a constant value of the load force the valve flow gain with valve displacement is from equation 15:

$$\frac{Q}{X} = \left(\frac{\partial Q}{\partial X}\right)_{P_m} = K_Q \sqrt{\left(1 - \frac{P_m}{P_S}\right)}$$

To develop the equations for the flow through each side of the valve with changes in the pressures P_1 and P_2 it is necessary to apply the method of small perturbations to the valve flow equations.

For the flow entering the actuator we have:

$$Q_1 = C_Q \pi dX \sqrt{\frac{2}{\rho}(P_S - P_1)} = KX\sqrt{(P_S - P_1)} \quad (16)$$

$$K = C_Q \pi d \sqrt{\frac{2}{\rho}}$$

Thus the flow and changes in the flow for small changes in X and P_1 are given by:

$$\Delta Q_1 = q_1 = \frac{\partial Q_1}{\partial X} x + \frac{\partial Q_1}{\partial P_1} p_1 \quad (17)$$

The lower case letters denote small changes in the variables.
From equation 16 we have:

$$\frac{\partial Q_1}{\partial P_1} = \frac{KX}{2\sqrt{(P_S - P_1)}}(-1) \text{ and } \frac{\partial Q_1}{\partial X} = K\sqrt{(P_S - P_1)} \quad (18)$$

Control System Design

Similarly for the flow leaving the actuator:

$$Q_2 = C_Q \pi dX \sqrt{\frac{2}{\rho} P_2} = KX\sqrt{P_2}$$

$$Q_2 = f(X, P_2), \quad \Delta Q_2 = q_2 = \frac{\partial Q_2}{\partial X} x + \frac{\partial Q_2}{\partial P_2} p_2 \quad (20)$$

$$\frac{\partial Q_2}{\partial P_2} = \frac{KX}{2\sqrt{P_2}} \quad \text{and} \quad \frac{\partial Q_2}{\partial X} = K\sqrt{P_2} \quad (21)$$

Thus it is seen that the values of the flow coefficients used to express the changes in the flows with changes in X and with the pressures are dependent on the steady state values of the pressures. This is why the equations are only applicable for small changes about a steady state condition.

The symmetry of the valve flow characteristics arising from the use of an equal area actuator results in the following:

$$\frac{\partial Q_1}{\partial X} = \frac{\partial Q_2}{\partial X} = C_x \quad (22)$$

And

$$\frac{\partial Q_1}{\partial P_1} = -\frac{\partial Q_2}{\partial P_2} = -C_p \quad (23)$$

Thus the changes in flow are given by:

$$q_1 = C_x x - C_p p_1 \quad (24)$$

and:

$$q_2 = C_x x + C_p p_2 \quad (25)$$

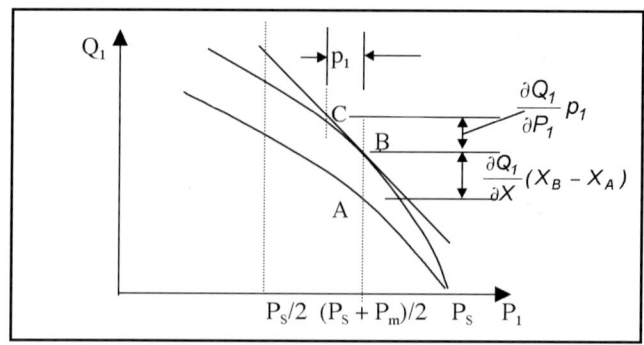

Figure 9. Valve flow coefficients

Figure 9 gives a representation of equation 24 where p_1 is the change in pressure away from the steady state value of $(P_s + P_m)/2$. For reasons of symmetry the changes in p_1 and p_2 will be equal in magnitude and opposite in sign i.e. $p_1 = -p_2$.

4.2. Actuator flows

The flow between the valve and the actuator on both sides of the piston will be subject to compressibility effects as described in section 3.

For an actuator velocity, u, the inflow is given by:

$$q_1 = C_x x - C_p p_1 = Au + \frac{V}{\beta}\frac{dp_1}{dt} = Au + \frac{V}{\beta}sp_1$$

$$\therefore p_1 = \frac{C_x x - Au}{C_p + \frac{V}{\beta}s} \quad (26)$$

And:

$$q_2 = C_x x + C_p p_2 = Au - \frac{V}{\beta}sp_2$$

$$\therefore p_2 = \frac{Au - C_x x}{C_p + \frac{V}{\beta}s} \quad (27)$$

The analysis is for equal volumes on either side of the piston which would apply to the piston in the mid-position for equal pipe volumes on both sides. It is

Control System Design

seen from equations 26 and 27 that $p_1 = -p_2$ because of the symmetry created by the equal area actuator.

4.3 Actuator force

This analysis is concerned with small changes about a steady state operating condition for an external force that is assumed to be constant so the analysis will only need to consider changes in force due to inertia and friction. Friction is usually considered to compose of coulomb, which is assumed to be unchanging, and a speed dependent element that is assumed to vary proportionally with the velocity. In practice the level of coulomb friction is known to vary considerably but in an unpredictable manner and this effect, consequently, is not included in this analysis.

Thus changes in the forces can be given by:

$$(p_1 - p_2)A = m\frac{du}{dt} + C_f u = msu + C_f u \qquad (28)$$

Substituting equations 26 and 27 into 28 gives:

$$2AC_x x = 2A^2 u + (C_p + \frac{V}{\beta}s)(ms + C_f)u$$

$$\frac{C_x}{A}x = (1 + \frac{C_p C_f}{2A^2} + (\frac{C_p m}{2A^2} + \frac{C_f V}{2A^2 \beta})s + \frac{mV}{2A^2 \beta}s^2)u$$

The transfer function between u and x is of the form:

$$\frac{u}{x} = \frac{C_x/A}{(1 + 2\frac{\zeta}{\omega_n}s + \frac{1}{\omega_n^2}s^2)} \qquad (29)$$

Where: $\omega_n = \sqrt{\frac{2\beta A^2}{Vm}}$ (30)

And: $\zeta = \frac{C_p}{2A}\sqrt{\frac{m\beta}{2V}} + \frac{C_f}{2A}\sqrt{\frac{V}{2m\beta}}$ (31)

This has assumed that the $\frac{C_f C_p}{2A^2}$ term is negligibly small.

4.4 Comments

Generally it is found that the value of ζ predicted by equation 31 is lower than in practical systems, some of the reasons for this include:

- The value of C_f is usually difficult to predict with any accuracy and actuator friction can vary considerably, particularly at low velocities.
- The electrohydraulic valve has been assumed to have an instantaneous response. In many applications the valve has a much higher response than the hydraulic natural frequency of the system but non-linearities at the null position, such as friction, leakage and coil hysteresis can affect the system performance.
- The value of C_p, for a given load pressure, varies over the flow range from zero at the valve null position to a maximum when the valve is fully open but in the analysis this has to be taken as a constant. In fact valves will leak over a range of valve positions around the null point as specified by the pressure gain quoted by the manufacturer which affects the system response and steady state accuracy. This is discussed in section 6.3.1.

4.5 Actuator position

Changes in actuator displacement are the integral of velocity that can be written as:

$$y = \frac{u}{s}$$

Thus, from equation 29 we get:

$$\frac{y}{x} = \frac{C_x / A}{s(1 + 2\frac{\zeta}{\omega_n}s + \frac{1}{\omega_n^2}s^2)} \quad (32)$$

Equation 32 is the open loop transfer function that relates changes in actuator displacement to small changes in the valve position with a constant external force acting on the actuator rod. This is a third order equation that can give rise to instability in a closed loop.

The effect of the load mass has been to add a second order equation to the first order system obtained in equations 2 and 3 because of the hydraulic system stiffness created by the fluid compressibility. The value of the flow coefficient C_x

is the same as that of the flow coefficient K_Q used in 2.1 for the simple system when the actuator is unloaded and $P_1 = P_2$.

4.6 Valve selection

Valves are usually rated on their flow, Q_R, with both outlet ports connected together at an overall pressure drop of 70 bar (ΔP_R). The flow at any other pressure difference is given by:

$$Q = Q_R \sqrt{\frac{\Delta P}{\Delta P_R}}$$

Thus for the analysis the overall pressure difference in the valve is = $2(P_s - P_1)$ or $2P_2$ where these pressures are determined from the load force acting on the actuator rod. The valve size needs to be slightly larger than the maximum required flow at the maximum valve input current.

4.7 Pressure shock control in open loop systems

In the open loop operation of valve actuator systems, fast opening of valves can cause considerable pressure shocks which may result in noise, excessive vibration of the load and, possibly, of the machine itself. Such shocks can be reduced by ramping the valve (open and/or closed) over a short period of time.

The variation in actuator displacement and pressure can be analysed by using the linearised equations 28 and 32 with the valve displacement changing with time at a constant rate (ramp). A comparison of the responses with reduced rates of change of valve movement will show that the amplitude of the pressure shocks will be reduced. The actuator position will also be reduced but not significantly, as the ramping time is usually of the order of 10 to 50 ms.

An example of the effect of the valve ramping time is given in chapter 11, 13.

5. Frequency response

The stability of a closed loop system can be obtained by examining the frequency response of the open loop system. For the simple system the open loop is an integrator, and in cases where the natural hydraulic frequency is high, the second order term in equation 32 can be neglected.

5.1 Simple actuator

When supplying flow into an actuator as a sinewave, the actuator displacement will also be a sinewave but displaced by 90^0 phase shift as shown in Figure 10. This is because for positive flow the actuator moves in one direction, returning to the start point during the negative flow period.

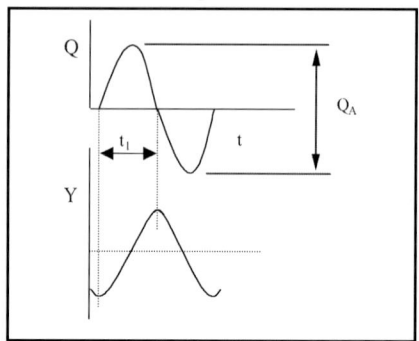

Figure 10. Sinusoidal flow variations and actuator displacement

Thus during the time period t_1 the fluid volume entering the actuator will be:

$$V = \int_0^{t_1} \frac{Q_A}{2} \sin \omega t \, dt = \frac{Q_A}{\omega} \text{ for } t_1 = \frac{\pi}{\omega}$$

For this volume the amplitude of the actuator displacement is

$$Y_A = \frac{V}{A} = \frac{Q_A}{A\omega} \qquad (33)$$

The actuator displacement is therefore inversely proportional to the frequency with a phase lag of 90^0. Mathematically this is obtained by replacing the 's' term by $j\omega$ in the transfer function. From equation 3 for the actuator this gives:

$$\frac{y}{x} = \frac{K_Q}{Aj\omega}; \quad \left|\frac{y}{x}\right| = \frac{K_Q}{A\omega} \; \& \; \phi = -90^0 \qquad (34)$$

It is preferable to plot the frequency response as a Bode diagram that uses logarithmic axes so that equation 34 becomes:

$$20 \log R = 20 \log(\frac{K_Q}{A}) - 20 \log \omega \qquad (35)$$

Control System Design

$R = \left|\dfrac{y}{x}\right|$, and 20logR is the amplitude in decibels (dB).

Equation 34 is a straight line having a slope of -20 dB/decade that crosses the 0dB axis when $\omega = K_Q/A$.

5.2 Valve actuator system

The frequency response of the open loop transfer function of equation 32 can be obtained by replacing the 's' terms with $j\omega$ which gives for the amplitude and phase angle:

$$R = \dfrac{C_x/A}{\omega\sqrt{(1-(\dfrac{\omega}{\omega_n})^2)^2 + (2\zeta\dfrac{\omega}{\omega_n})^2}} \qquad (36)$$

$$20\log R = 20\log(\dfrac{C_x}{A}) - 20\log\omega - 20\log\left((1-(\dfrac{\omega}{\omega_n})^2)^2 + (2\zeta\dfrac{\omega}{\omega_n})^2\right)^{0.5} \qquad (37)$$

(37)

$$\phi = -\tan^{-1}\left(\dfrac{2\zeta\dfrac{\omega}{\omega_n}}{(1-(\dfrac{\omega}{\omega_n})^2)}\right) - 90^0 \qquad (38)$$

The variations of the amplitude ratio and phase angle with frequency are shown in Figure 11.

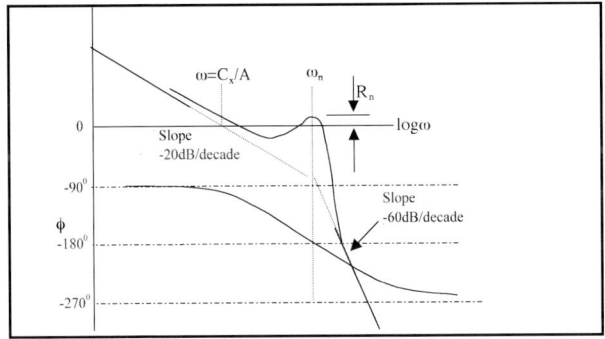

Figure 11. Bode plot for valve actuator open loop transfer function.

The amplitude ratio of equation 37 can be simplified by approximating the value of the second order term for $\omega < \omega_n$ and for $\omega > \omega_n$.

Thus for $\omega < \omega_n$: $20 \log R \rightarrow 20 \log (\frac{C_x}{A}) - 20 \log \omega$ because the powers of $\frac{\omega}{\omega_n} \ll 1$ so that the second order term in the denominator tends to unity.

And for $w > w_n$: $20 \log R \rightarrow 20 \log (\frac{C_x}{A}) - 20 \log w - 40 \log (\frac{\omega}{\omega_n})$ when the second order term in the denominator tends to $\frac{\omega}{\omega_n}$. This is a straight line having a slope of -60 dB/decade.

These straight-line approximations, or asymptotes, are shown dotted in Figure 11 and the actual response deviates from these in a small range of frequencies around the natural frequency.

From equation 36, at the natural frequency:

- the amplitude ratio $R_n = \dfrac{C_x/A}{2\zeta\omega_n}$ (39)

- the phase angle is -180^0.

6. Stability of the closed loop position control system

6.1 Stability criterion

For stability of the closed loop system, the open loop response transfer function must have an amplitude ratio that is less than unity when the phase lag is 180^0. Thus for the electrohydraulic position control system it will be necessary to set the power amplifier gain to a value that gives an open loop amplitude ratio that is less than unity. In order to provide some margin it is usual to design the system so that the gain is 0.5, or -6 dB. The design should also aim to achieve a phase lag of less than around 140^0 when the amplitude ratio is unity (0 dB) i.e. a phase margin of at least 40^0.

6.2 System design

The system frequency response analysis in 5.2 has shown that -180^0 phase lag occurs at the natural frequency when the amplitude ratio of the valve actuator system is $\dfrac{C_x/A}{2\zeta\omega_n}$.

The block diagram of Figure 4 is shown modified in Figure 12 to include the effect of the fluid compressibility using the transfer function for the valve actuator system from equation 32.

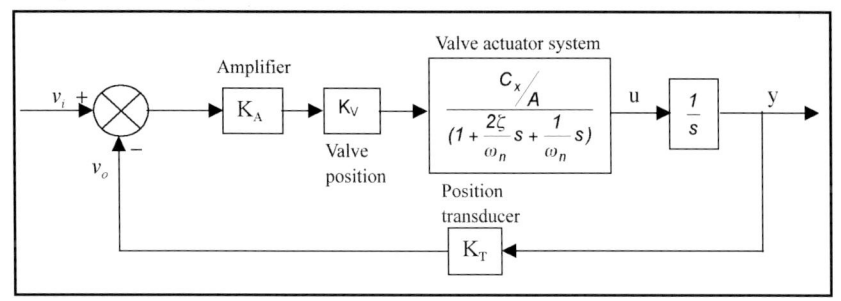

Figure 12. Electrohydraulic position control system block diagram.

From equation (39), which gives the amplitude ratio at the natural frequency when $\phi = -180^0$ the stability criterion for the system in Figure 12 requires that:

$$\dfrac{C_x K_A K_V K_T}{2\zeta\omega_n A} = \dfrac{K_L}{2\zeta\omega_n} \le \dfrac{1}{2} \qquad (40)$$

Where K_L is the open loop gain which $= K_A K_V K_T \dfrac{C_x}{A}$.

The value of the system open loop gain required for stability can be obtained from equation 40 where it is seen that for this to be satisfied $K_L \le \zeta\omega_n$.

Typically, taking a value for ζ of 0.2, means that K_L needs to be $\le \dfrac{\omega_n}{5}$. Consequently, having determined the hydraulic natural frequency enables an initial value of the system gain to be obtained. Adjustment of the amplifier gain K_A can be used to alter the system gain so as to provide a level that satisfies the stability criteria.

The valve gain, K_V, can be established by selecting one having the appropriate flow rating to achieve the required actuator velocity and having the desired frequency response. The valve performance is normally given in terms of the flow variation with input current at a given pressure drop so that the valve flow gain $= K_V C_X \{(m^3/s)/A\}$.

In this analysis the valve has been assumed to respond instantaneously. The valve dynamic performance is usually described by its frequency response and if this produces an insignificant phase lag at the system hydraulic natural frequency then the valve can be treated as a steady state gain. The valve manufacturer will show how the valve phase lag and amplitude ratio vary with frequency and this information can be used if necessary in the system Bode plot.

The feedback transducer gain, K_T, is usually selected to have a voltage range of $\pm 10V$ for the required actuator movements about the mid-position. Thus, knowing the required valve input current enables an amplifier gain, K_A, to be selected that will give the value of K_L that provides the necessary gain and phase margins.

6.3 Steady state accuracy

For the position control system in Figure 12, the closed loop transfer function is given by:

$$\frac{v_0}{v_i} = \frac{K_L}{K_L + s(1 + \frac{2\zeta}{\omega_n}s + \frac{1}{\omega_n^2}s^2)} \tag{41}$$

In the steady state the 's' terms are zero and from equation (41) we get $V_0 = V_i$ or $y = \frac{V_i}{K_T}$.

The error signal, $V_i - V_0 = 0$ because the control valve has to be closed for the actuator to be at rest. Mathematically there needs to be a zero input to the integral action of actuator velocity to displacement for steady state conditions to be obtained.

The linear analysis used to develop the open loop transfer function excluded non-linear effects such as leakage through the valve at the null position, electrical and mechanical hysteresis in the valve and actuator friction forces (coulomb and stiction). All of these affect the steady state accuracy.

The block diagram in Figure 13 shows pressure and force from equations 26, 27 and 28 for an equal area actuator with the piston in the central position when the perturbation pressure $P = P_1 = -P_2$. The diagram also includes changes in the external load force, f_E, as input to the system. For $f_E = 0$, the transfer function for y to changes in x is the same as that given in equation 32 and shown in Figure 12.

For the closed loop position system of Figure 12 the effect of changes in external force can be evaluated from Figure 13 by considering, for example, a fixed input voltage when v_i will be zero. It must be remembered that the system

Control System Design   157

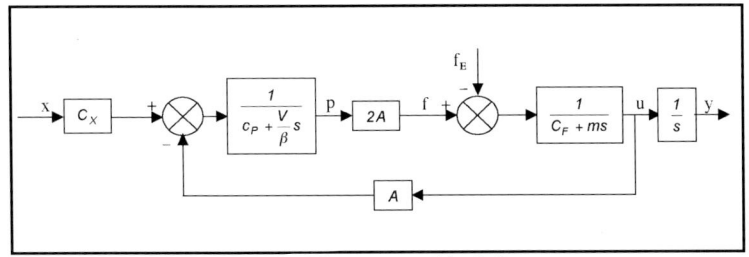

Figure 13. Valve actuator block diagram

transfer function applies only for **changes** in the variables about a steady state condition where the initial values are known.

As has been described, the frequency response of the linearised system allows the system stability to be evaluated. However, it is necessary to examine the effects that the non-linear elements have on the steady state accuracy of the system.

6.3.1 Valve leakage

The linearisation of the valve characteristics enabled the small perturbation technique to be used to obtain the variation in flow for small changes in pressure as represented by the flow coefficient C_p. For the ideal valve that closes completely (i.e. $x = 0$) in the null position the value of C_p is zero which would cause very low damping and probably require a low system gain to obtain stability of the closed loop system.

Conversely, for high load forces when the value of $(P_S - P_l)$ approaches zero the value of C_p will approach infinity. These variations in C_p are a major limitation of the linearisation technique but in reality the valve will leak at the null position and it may even be deliberately underlapped to improve its flow characteristics. Consequently, the effect of leakage through the valve for positions close to the null can be evaluated by treating the valve as underlapped as shown in Figure 14.

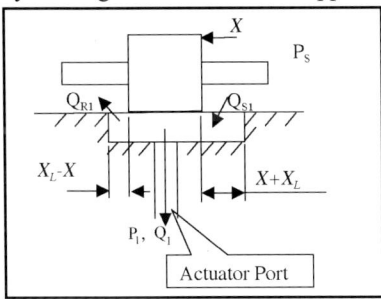

Figure 14. Valve underlap

Underlapping results in the port width being greater than the width of the spool land so that some fluid passes from the port to the return (Q_{R1} in Figure 13) for $0 < X < X_L$. When $X > X_L$, the valve behaves as a zero lap valve.

The flows through the valve can be analysed as follows:

$$Q_1 = Q_{S1} - Q_{R1} = K(X + X_L)\sqrt{P_S - P_1} - K(X_L - X)\sqrt{P_1} \quad (42)$$

Now:

$$C_{X1} = \frac{\partial Q_1}{\partial X} = K\sqrt{P_S - P_1} + K\sqrt{P_1}$$

$$\text{for } P_1 = P_S/2, \; C_{X1} = 2K\sqrt{\frac{P_S}{2}} \quad (43)$$

and:

$$C_{p1} = \frac{\partial Q_1}{\partial P_1} = -\frac{K(X + X_L)}{2\sqrt{P_S - P_1}} - \frac{K(X_L - X)}{2\sqrt{P_1}}$$

$$\text{for } P_1 = P_S/2 \text{ and } X = 0, \; C_{p1} = \frac{-K}{\sqrt{P_S/2}} X_L \quad (44)$$

It can be seen from equation (42) that for a given valve opening X, for values of P_1 in the range from zero to the supply pressure the flow in the actuator port will change direction which is reflected in the value of C_{X1}. Also it is seen from equation (44) that the magnitude of C_{p1} is greater than zero when $X = 0$ which can be an advantage in some applications.

Equation 42 can be used to obtain the variation of P_1 with X for zero port flow (e.g. ports blocked). The same approach can be used to obtain the variation of P_2 with X so giving the change in actuator load pressure as the valve is moved through the leakage range as shown in Figure 15.

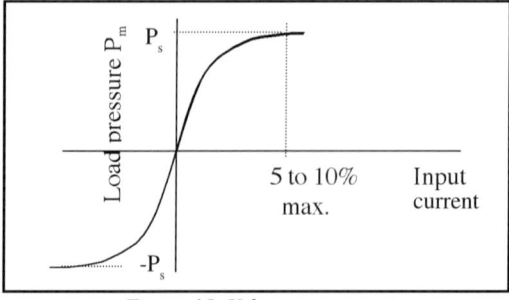

Figure 15. Valve pressure gain

Control System Design

This information is obtained from tests by the valve manufacturer and is referred to as the pressure gain. It can be seen therefore, that in the closed loop position control system, variations in the load pressure will require changes in the valve position in order to provide the requisite actuator load pressure. This will result in $V_i \neq V_o$ which means that the load force will affect the steady state accuracy.

The perturbation analysis can be applied to the underlapped valve in order to provide an indication of the effect on steady state performance.

Now for the valve: $q_1 = C_{xI}x + C_{pI}p$. For steady state conditions, $q_1 = 0$ so $C_{xI}x = -C_{pI}p_I$

And from equations 43 and 44
$$p_1 = -\frac{C_{x1}}{C_{p1}}x = \frac{P_s}{X_L}x \qquad (45)$$

{for $x = 0$ (null position) & $P_1 = \dfrac{P_s}{2}$ (zero load force)}

From symmetry of the valve characteristics the change in the pressure P_2 is:

$$p_1 = p = -p_2 \qquad (46)$$

Consequently, the change in the actuator force, f, is given by:

$$f = 2pA \qquad (47)$$

Substituting equations 45 into 47 gives:

$$f = 2AP_s \frac{x}{X_L} \qquad (48)$$

The block diagram in Figure 16 shows the parameters involved in the steady-state operation of the closed loop position system when the actuator force change, f, must be equal to the change in the external force, f_E. The change in the actuator force is created by a change in the output position, Y.

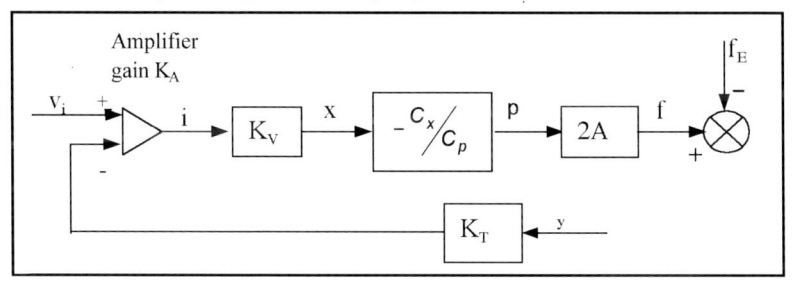

Figure 16. Block Diagram of steady state conditions

For zero changes in the input command signal, (i.e. $v_i = 0$) the corresponding change in x is given by:

$$x = -K_T K_A K_V y \qquad (49)$$

Thus, from equation 48 we get:

$$f = f_E = -2AP_S \frac{K_T K_A K_V y}{X_L}$$

Or the system stiffness is:

$$\frac{f_E}{y} = -\frac{2AP_S K_T K_A K_V}{X_L} \qquad (50)$$

This linear analysis is useful to illustrate the effect of system parameters on the system stiffness, but is rather idealised in that it takes no account of valve hysteresis. Because of the linearisation the value of the stiffness only applies to small changes in X and the pressures P_1 and P_2 about any steady state operating condition although the valve characteristics are very linear in the underlapped region.

This effect is referred to as the stiffness of the closed loop system and to increase this, and, consequently reduce the steady state error, the system gain needs to be increased. However, increases in the system gain are limited by the available gain margin and additional methods are sometimes required to avoid this problem.

6.3.2 Valve hysteresis

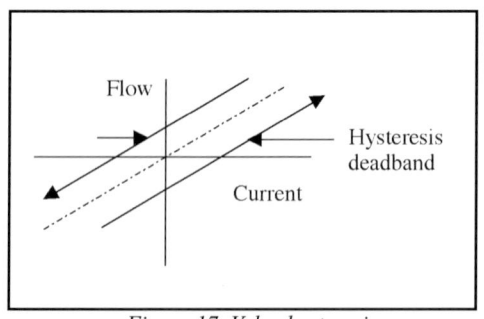

Figure 17. Valve hysteresis

The magnetic hysteresis in the valve solenoids and friction in the components will affect the valve position for a given input current as shown in Figure 17. As in section 6.3.1, increasing the system gain will reduce the error voltage required to provide the necessary valve current to hold the valve in the appropriate position.

7. The improvement of closed loop system performance

In addition to controlling actuator position, electrohydraulic systems are used to control output variables that include:

- Actuator velocity
- Actuator force

7.1 Position control

As was shown in section 6, the steady state accuracy of a position control system is limited by the non-linear characteristics inherent in the components. Increasing the system open loop gain can reduce these effects but this action is limited by stability considerations. The same limitation naturally arises if it is desired to increase the system response by increasing the system gain.

To improve steady state accuracy and/or system dynamic performance, compensation networks can be applied to the electronic control system and some of the available methods will be described.

The addition of an integrating amplifier in the forward path of the position control system would eliminate steady state errors but the consequent introduction of 90^0 phase lag in the integrator will make the position control system unstable. To avoid this the compensation system can have an integrator that is added to the proportional control element. This will be discussed in a later section.

7.2 Velocity control

The use of a velocity transducer to provide a feedback signal in the place of actuator position will provide a velocity control system. The closed loop transfer function obtained from Figure 12 when using velocity feedback with a velocity transducer gain, K_T, is given by:

$$\frac{u}{v_i} = \frac{K_F}{(1 + \frac{2\zeta}{\omega_n}s + \frac{1}{\omega_n^2}s^2) + K_L} \tag{52}$$

Where $\quad K_F = K_A K_V \dfrac{C_X}{A} = \dfrac{K_L}{K_T}$

In the steady state we get:

$$\frac{u}{v_i} = \frac{K_F}{1+K_L} = \frac{K_L}{K_T(1+K_L)}$$

If the error voltage is zero, $u = \dfrac{v_i}{K_T}$ but this can be obtained only if $K_L \gg 1$. This is because the valve has to be open in order to pass the necessary flow. As the open loop gain is increased, the required error voltage to produce a given valve displacement will be reduced.

In this system, which is second order, an integrating amplifier can be placed in the forward path so that the closed loop transfer function of equation 52 becomes:

$$\frac{u}{v_i} = \frac{K_F}{s\left(1 + \dfrac{2\zeta}{\omega_n}s + \dfrac{1}{\omega_n^2}s^2\right) + K_L} \qquad (53)$$

For steady state conditions, following a step input in v_1, the voltage error signal supplied to the integrator will be zero, the integrator output having the level appropriate to provide the required valve flow. Putting the 's' terms to zero in equation 53 produces the same result. This system is now of third order and therefore its stability margin will have to be determined in the same manner as that used for the position control.

7.3 Pressure control

In systems where it is required to control the actuator force, a pressure transducer can be used to provide a feedback signal so that the system is closed loop on actuator pressure instead of its position. This will provide actuator force control so that the actuator will be displaced such that the force is kept constant. Clearly this process will be affected by the characteristics of the load. This type of control is frequently used in material testing machines where it is desired to vary the force in a controlled manner and the actuator displacement will then depend on the mechanical stiffness of the test component.

Control System Design

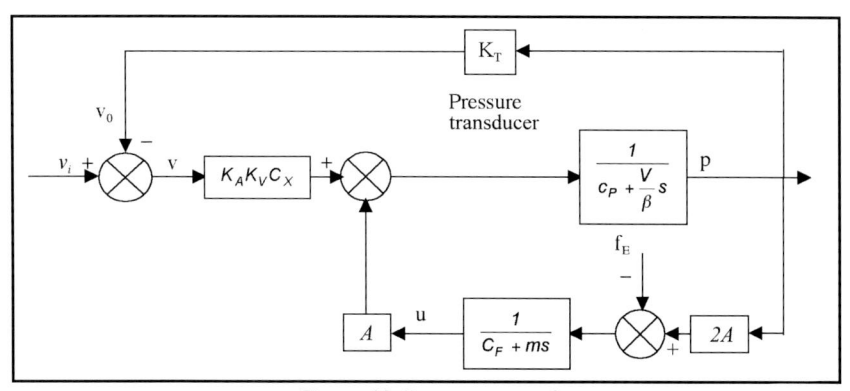

Figure 18. Pressure control

From Figure 13 the block diagram for the pressure control system is shown in Figure 18 where the pressure is fed back by the pressure transducer having a gain K_T.

From Figure 18 with $f_E = 0$ (constant load force), we get:

$$\frac{p}{v} = \frac{K_A K_V C_X (C_f + ms)}{2A^2 + (C_P + \frac{V}{\beta} s)(C_f + ms)} \qquad (54)$$

For the closed loop system the transfer function is given by:

$$\frac{p}{v_i} = \frac{K_A K_V C_X (C_f + ms)}{2A^2 + (C_P + \frac{V}{\beta} s)(C_f + ms) + K_A K_V C_X K_T (C_f + ms)} \qquad (55)$$

The steady state gain is given by:

$$\frac{p}{v_i} = \frac{K_A K_V C_X C_f}{2A^2 + C_P C_f + K_A K_V C_X C_f K_T} \qquad (56)$$

For this situation, where the load force is constant, the actuator is moving at a velocity such that the viscous friction (coefficient C_f) absorbs the actuator force from the pressure increase. Clearly if this coefficient is zero then the actuator will be uncontrolled because there will not be any increase in the pressure. Pressure control can be combined with position control so that, in a press system, it will only be operational when the pressing action takes place.

In the fatigue test systems mentioned above the load force will increase with actuator displacement due to the component mechanical stiffness and the steady state open loop gain will be:

$$\frac{p}{v_i} = \frac{K_A K_V C_X}{C_P + K_A K_V C_X K_T}$$

For this situation the actuator displacement will increase to a level such that the increased pressure force equals the load force when the actuator velocity, u, will be zero. As with the position control, this steady state gain will vary due to the system non-linear characteristics and improvements can be obtained by the use of integral plus proportional control.

8. Compensation techniques

The objective of most compensation systems is to provide a high gain at low frequency in order to minimise steady state errors whilst modifying the system frequency response so as to obtain the requisite gain and phase margins. Some of the available techniques are described.

8.1 Integral plus proportional compensation

The transfer function for this element is given by:

$$K_p + \frac{K_I}{s} = \frac{K_p s + K_I}{s} = \frac{K_I}{s}(1 + \frac{K_P}{K_I}s) \qquad (57)$$

For frequency inputs the amplitude ratio and phase change are given by:

$$R = \frac{K_I \sqrt{1 + (\frac{K_P}{K_I}\omega)^2}}{\omega} \qquad (59)$$

$$\phi = -90° + \tan^{-1}\left(\frac{K_P \omega}{K_I}\right) \qquad (60)$$

Control System Design

Here when $\omega \gg \dfrac{K_1}{K_P}; R \to K_P$

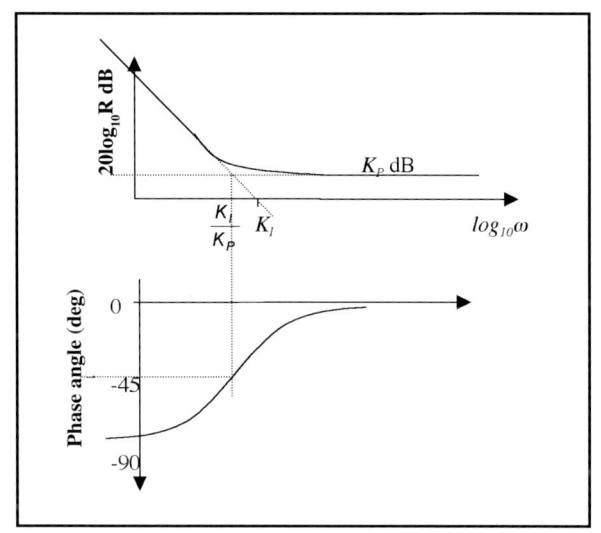

Figure 19. Frequency response for integral plus proportional control

The frequency response is shown in Figure 19 where at low frequencies (when $\omega \to 0$), $R \to \dfrac{K_I}{\omega}$ which increases with reducing frequency (i.e. has the characteristics of an integrator) so eliminating steady state errors. The ratio $\dfrac{K_I}{K_P}$ needs to be selected in relation to the hydraulic natural frequency and the system frequency response as shown in Figure 11.

If this ratio is too low the system response will be slow and if it is too high it will not have the required effect. The system gains can be selected so as to provide the requisite gain and phase margins.

8.2 Proportional plus derivative control

The transfer function for this compensator is given by:

$$K_P + K_D s = K_P \left(1 + \dfrac{K_D}{K_P} s\right) \qquad (61)$$

This has a frequency response that is given by:

$$R = K_P \sqrt{\left(1 + \left(\frac{K_D}{K_P}\omega\right)^2\right)} \tag{62}$$

$$\phi = \tan^{-1}\left(\frac{K_D}{K_P}\omega\right) \tag{63}$$

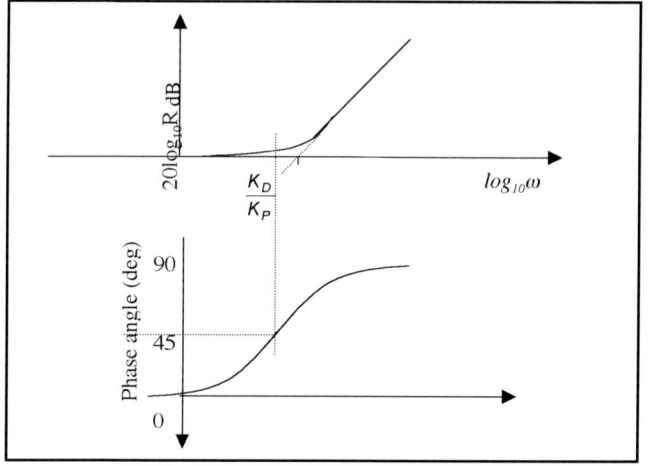

Figure 20. Frequency response for proportional plus derivative control

As can be seen from Figure 20 this compensator provides a phase advance characteristic in that the phase angle is positive which will reduce the overall phase lag in the system. However, the amplitude ratio also increases and is zero at low frequencies so there is no effect on the steady state gain. A major problem with derivative control is that it will amplify noisy input signals and for this reason it is often not used. It is more common to use a phase advance compensator, which is described, in the next section.

8.3 Phase advance compensation

As an alternative to derivative control phase advance can be provided by a transfer function of the form:

$$\frac{1 + \alpha\tau s}{1 + \tau s} \tag{64}$$

Control System Design

By putting $\alpha > 1$ the magnitude of the phase angle of the numerator is always greater than that of the denominator and the overall phase angle is zero at high and low frequencies.

Thus:

$$R = \sqrt{\frac{1 + (\alpha\tau\omega)^2}{1 + (\tau\omega)^2}} \tag{65}$$

and

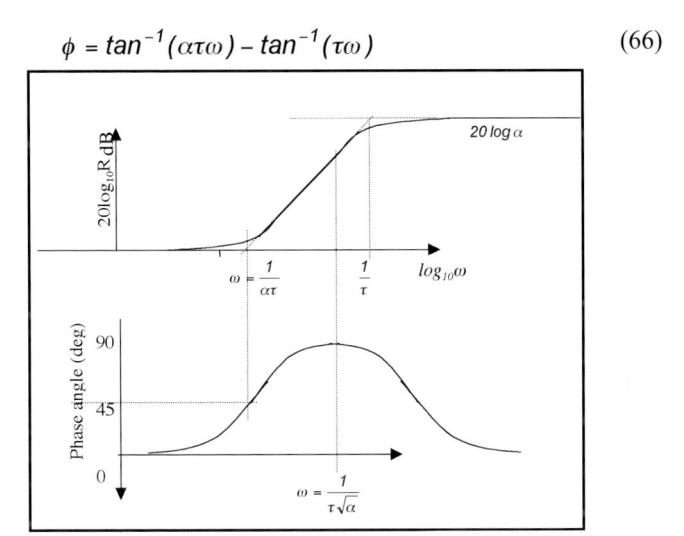

$$\phi = \tan^{-1}(\alpha\tau\omega) - \tan^{-1}(\tau\omega) \tag{66}$$

Figure 21. Phase advance frequency response

As shown in Figure 21 the phase advance system provides maximum phase advance at a frequency that is determined by the selection of the values of the coefficients α and τ so as to stabilise an unstable system and/or increase the natural frequency so improving its response.

8.4 Proportional, Integral and Derivative (PID) control

There is a general approach to compensation, which is referred to as Proportional, Integral and Derivative Compensation in the Forward Path, or PID. This can be represented by the following transfer function:

$$K_P + \frac{K_I}{s} + K_D \left(\frac{1 + \alpha\tau s}{1 + \tau s} \right)$$

The frequency response of the compensator can be modified by adjusting the values of:

1. The proportional gain coefficient K_p
2. The integral gain coefficient K_I
3. The derivative gain coefficient K_D
4. The time constant τ and phase advance term α.

The electronic control amplifiers for proportional and servo valves usually contain an adjustable PID system. The application of PID, because of the large combination of adjustable coefficients that are available, is difficult to discuss in a generalised fashion and it is often 'tuned' after it has been installed on the system.

8.5 Pressure feedback

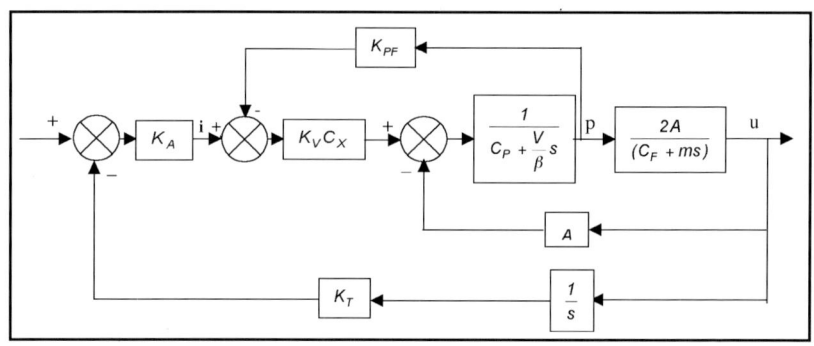

Figure 22. Pressure feedback control ($f_E = 0$)

Pressure feedback has been discussed as a means for providing force control but it can also be fed back inside a position control loop for increasing the damping ratio of the open loop system as shown in Figure 22. Obtaining the open loop transfer function can show this effect. Thus:

$$p = \left(\frac{1}{C_P + \frac{V}{\beta}s} \right) \left[(i - K_{PF}p)K_V C_X - Au \right]$$

Rearranging this equation gives:

$$\left[(C_P + \frac{V}{\beta}s) + K_{PF}K_V C_X \right] p = K_V C_X i - Au \qquad (67)$$

Control System Design 169

Also:
$$\frac{u}{p} = \frac{2A}{C_f + ms} \quad (68)$$

Substituting for p from equation 68 into equation 67 gives:

$$\left[(C_P + \frac{V}{\beta}s) + K_{PF}K_V C_X\right]\frac{(C_f + ms)}{2A}u = K_V C_X i - Au$$

And so:

$$\left(\frac{mV}{2A^2\beta}s^2 + (\frac{mC_P}{2A^2} + \frac{C_f V}{2A^2} + \frac{K_{PF}K_V C_X m}{2A^2})s + 1\right)u = \frac{K_V C_X}{A}i \quad (69)$$

The term $\frac{C_j C_f}{2A^2}$ has been neglected as being $\ll 1$.

On comparing equation 69 with that for the damping factor in equation 31 it is seen that the pressure feedback has introduced the term $\frac{K_{PF}K_V C_X m}{2A^2}$ which increases the value of the damping factor and thus reduces the resonance at the natural frequency in the Bode plot of Figure 11. This can provide a considerable benefit in some systems where the damping factor is low (typically < 0.2) which may allow the system gain to be increased whilst maintaining an adequate stability margin.

A problem that can arise with pressure feedback is that it reduces the system stiffness and, consequently, increases the steady state error as discussed in section 6.3.1. The use of a high pass filter can prevent this problem as it excludes all signals below a set frequency. This will, therefore, only pass the feedback pressure transducer signal when transient changes occur and not when the system is in the steady state.

9. System frequency response tests

As has been described, frequency response analysis can be used for the design of the electrohydraulic closed loop system. For testing a physical system it is clear that if the gain is set at too high a value the closed loop system will be unstable, consequently, during commissioning the gain should be set at a low value so that frequency response tests can be carried out.

Frequency response tests on closed loop systems can provide an open loop Bode plot by the use of a Nichols Chart that converts closed to open loop and open to closed loop performance. However, the problem with closed loop testing is that the input signal to the valve will vary with the frequency and it is normally desired to keep this constant because of non-linearities in the valve.

Open loop tests can be carried out by removing the feedback signal but the main difficulties involved in this are:

- The actuator position may drift slightly if there is any bias in the frequency input signal and eventually the actuator may hit the end stops.
- At low frequencies the actuator movement may be longer than the available stroke when it will be necessary to reduce the input signal amplitude at low frequencies. At high frequencies it may be necessary to increase the input signal in order to obtain actuator displacement amplitudes that can be measured with sufficient accuracy. This will, however, mean that the input amplitude to the valve is not constant for all frequencies.

The actuator mountings may introduce a low stiffness in the mechanical system (natural frequency ω_m) including the rod connection to the load, which can have an effect on the system dynamic performance. The natural frequency, ω_s, of the composite system is:

$$\omega_s = \frac{\omega_n \omega_m}{\sqrt{\omega_n^2 + \omega_m^2}} = \frac{\omega_n}{\sqrt{1 + \left(\frac{\omega_n}{\omega_m}\right)^2}} \qquad (70)$$

This can affect the system stability which may depend on the location of the position transducer i.e. actuator rod to actuator cylinder or load to machine base. It can be seen from equation 70 that the system natural frequency, ω_s, is significantly reduced if the two frequencies are equal or when $\omega_m < \omega_n$ and, consequently, it is recommended that the mechanical stiffness should be around five to ten times that of the hydraulic system.

10. Pump controlled systems

The use of valve control creates restrictive losses in hydraulic systems that are dissipated in the fluid as heat and can lead to low power transmission efficiencies. Low system efficiency increases the size of the prime mover and that of the cooling system. The advantage of the method over other forms of control is its compactness, high natural frequency and flexibility.

Control System Design   171

However, as an alternative, the pump can be directly connected to the actuator as a hydrostatic system thus improving the power transmission efficiency to a theoretical level of 100%. The pump displacement is controlled by an electrohydraulic servo valve that operates in the closed loop system in the same way as in the valve controlled system i.e. from the position voltage error signal.

Figure 23 shows the hydraulic circuit for a variable displacement pump driving an equal area actuator. As with rotary hydrostatic systems, a boost make-up flow is required to compensate for leakage losses in the pump. In this system,

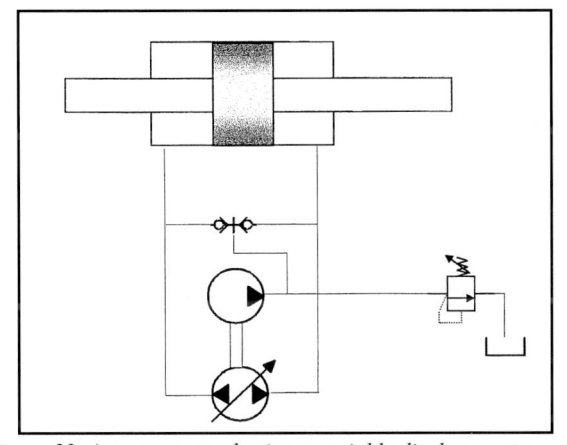

Figure 23. Actuator control using a variable displacement pump

therefore, only one side of the actuator is pressurised, if the load force changes direction, the opposite side of the pump becomes pressurised and the appropriate boost check valve is opened.

Pump controlled systems reduce the cooling requirement and the running cost and in some systems can be an attractive alternative to valve control. However, there are some major disadvantages that include:

a)   The pump has to be dedicated to a single actuator, or a mechanically grouped set of actuators. This can increase the capital cost in some installations but this must be balanced against reduced operating costs.

b)   The pressurised volumes are likely to be larger than those in valve controlled systems that will reduce the hydraulic natural frequency. However, in many actuator systems, the volume of the actuator dominates the hydraulic stiffness.

c)   The frequency response of the pump displacement controller may be lower than that of an electrohydraulic control valve which could reduce the available gain margin in the closed loop system. However, taking into consideration point b) concerning the hydraulic natural frequency this may not be a problem.

The transfer function of the hydraulic system in that controlled by a pump is similar to that given in equation 32 for the servo valve control where C_x is the flow gain of the pump for changes in the pump displacement controller position x.

CHAPTER ELEVEN

PERFORMANCE ANALYSIS

11. PERFORMANCE ANALYSIS

1. Introduction

This chapter is devoted to the examination of case studies in order to illustrate the various aspects that are involved in the analytical design of fluid power systems and the evaluation of component performance.

2. Meter-in control

It is required to select an appropriate valve for an actuator having a 50 mm piston and 28 mm rod diameter that is supplied from a fixed displacement pump of 35 L/min capacity with a relief valve setting of 150 bar. The required actuator velocity is 0.2 m /s with an actuator force of 10 kN.

$$\text{Piston area } (A_P) = \frac{\pi \, 0.05^2}{4} = 1.96 \times 10^{-3} \, m^2$$

$$\text{Pump flow } (Q_P) = \frac{35}{1000} \frac{1}{60} = 5.8 \times 10^{-4} \, m^3 \, s^{-1}$$

$$\text{Max. velocity } (U_e) = \frac{Q_P}{A_P} = 0.3 \, m \, s^{-1}$$

For an actuator force of 10 kN, the piston pressure (P_p) will be given by:

$$P_p = \frac{10 \times 10^3}{1.96 \times 10^{-3}} = 51 \times 10^5 \, N/m^2 = 51 \, bar$$

For the relief valve set at 150 bar, the pressure drop across the restrictor (P_v) will be:

P_v = 150 - 51 = 99 bar

For an actuator velocity of 0.2 m s^{-1}:

The supply flow

$$Q_s = \frac{0.2}{0.3} \times 35 = 23.3 \text{ L/min} = 23.3 \times \frac{10^{-3}}{60} = 3.9 \times 10^{-4} m^3 s^{-1}$$

Re-arranging the orifice equation, the restrictor area (A) can be found from:

$$Q_s = C_q A \sqrt{\frac{2 P_v}{\rho}}$$

$$\therefore A = \frac{Q_s}{C_q \sqrt{\frac{2 P_v}{\rho}}}$$

Therefore, assuming the flow coefficient is 0.65 and the fluid density is 870 kg m^3, the restrictor area, A, is given by:

$$A = \frac{3.9 \times 10^{-4}}{0.65 \sqrt{\frac{2 \times 99 \times 10^5}{870}}} = 4 \times 10^{-6} m^2$$

This is equivalent to a restriction diameter of 2.25 mm. Alternatively, obtain a restrictor valve of an appropriate size from manufacturers' literature.

For a restrictor pressure drop of 7 bar which applies to the flow characteristics of an available adjustable restrictor valve in Table 1 at a flow of 23.3 L/min at 99 bar, the restrictor will pass a flow of:

$$Q = 23.3 \sqrt{\frac{7}{99}} = 6.2 \text{ L min}^{-1} @ 7 \text{ bar}$$

From Table 1 it can be seen that the restrictor setting requires approximately 3 turns.

Performance Analysis

Turns	Flow L/min
0	0
1	0.9
2	2.8
3	5.6
4	8.5
5	10.5

Table 1

2.1 The effect of load force changes

If the force increases to 20 kN, the required load pressure increases to 102 bar, thus reducing the available restrictor pressure drop to 48 bar. Hence the flow will reduce accordingly.
Thus:

$$Q_s = 23.3 \sqrt{\frac{48}{99}} = 16.2 \, L \, min^{-1}$$

For this flow the actuator velocity will be 0.14 m s^{-1}.

3. Valve control of a single ended actuator

A single ended actuator is to operate against a load that causes a pulling force on the actuator rod. For the given data determine the suitability of a proportional valve for controlling the actuator to give a retract velocity of 0.25 m/s. Also calculate the velocity when extending and the maximum pulling force capability of the actuator.

3.1 Data

Piston diameter   50mm
Rod diameter   25mm
Supply pressure   75 bar
Pulling force   3050 N
Valve rated flow   40 L/min   Valve rated pressure drop/metering land   35 bar
Piston area = 1.96 x 10^{-3} m^2
Annulus area = 1.47 x 10^{-3} m^2
Area ratio $\propto$ = 1.33

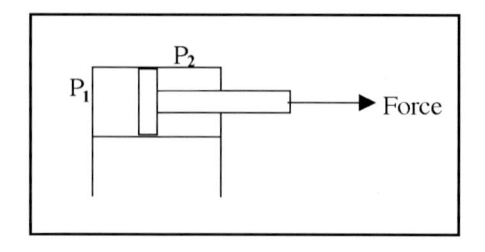

3.2 Actuator retraction

For this application, the force is negative and from equations (5) and (6), chapter 7, 5.4, for the actuator retracting we get:

$$P_1 = P_s \alpha^2 \left(\frac{1+\alpha R}{1+\alpha^3}\right)$$

$$\frac{P_2}{P_s} = \frac{\alpha(\alpha^2 - R)}{(1+\alpha^3)}$$

Stall force $F_s = P_s A_p = 14700\ N$ and so for $F = -3050\ N$ we get $R = -0.21$. This gives:

$P_1 = 28.5$ bar, $P_2 = 59$ bar and the valve pressure drop from the supply to the annulus during retraction is $\Delta P_2 = 75 - 59 = 16\ bar$

$Q_2 = 0.25 \times A_2 = 0.25 \times 1.47 \times 10^{-3}\ m^3/s = 3.7 \times 10^{-4}\ m^3/s = 22L/min$

Now for the valve, the rated flow at maximum opening is $Q_R = k_V \sqrt{\Delta P_R}$

For the actuator $\dfrac{Q_2}{\sqrt{\Delta P_2}} = \dfrac{3.7 \times 10^{-4}}{\sqrt{16 \times 10^5}} = 2.93 \times 10^{-7}\ \dfrac{m^3/s}{N/m^2}$

The valve data gives a value of

$$\frac{Q_R}{\sqrt{\Delta P_R}} = \frac{40}{6 \times 10^5 \times \sqrt{35 \times 10^5}} = 3.56 \times 10^{-7}\ \frac{m^3/s}{N/m^2} = k_V$$

The valve is therefore adequate to provide the require retract velocity.

3.3 Actuator extension

From chapter 7, 5.4 using equations 3 and 4, the pressures, are given by:

$$P_1 = P_S \left(\frac{1 + R\alpha^3}{1 + \alpha^3} \right)$$

$$P_2 = P_S \alpha \left(\frac{1 - R}{1 + \alpha^3} \right)$$

For R = - 0.21, these give:

$P_2 = 36\,bar$ and $P_1 = 11.3\,bar \therefore \Delta P_{s1} = 39\,bar$

The ratio of the extend and retract velocities is given by:

$$\frac{U_R}{U_E} = \sqrt{\frac{(P_S - P_2)_{retract}}{(P_2)_{extend}}} = \sqrt{\frac{16}{39}} = 0.64 \text{ and } U_E = 0.39\,m/s$$

The maximum force capability during extension is that which avoids cavitation of the piston end of the actuator. This is given by:

$$R = -\frac{1}{\alpha^3} = -\frac{1}{1{,}33^3} = -0.425$$

The maximum force, $F_{max} = -0{,}425 \times 14700 = -6247.5\,N$

4. Winch Application

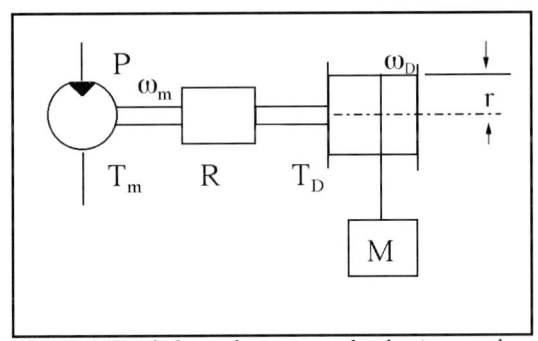

Figure 1 Winch driven by motor and reduction gearbox

4.1 Lifting the load

Winch drum torque $T_D = \dfrac{M g r}{\eta_d}$

Motor torque $T_m = \dfrac{T_d}{R \eta_R} = \dfrac{M g r}{R \eta_R \eta_d}$

Motor pressure $P = \dfrac{T_m}{D \eta_m} = \dfrac{M g r}{R D \eta_m \eta_R \eta_d}$

η_m = motor mechanical efficiency
η_R = reduction gearbox mechanical efficiency
η_D = winch drum mechanical efficiency

4.2 Lowering the load

$T_D = M g r \eta_D$ $\qquad T_m = \dfrac{T_d \eta_R}{R} = \dfrac{M g r \eta_R \eta_D}{R}$

$P = \dfrac{T_m \eta_m}{D} = \dfrac{M g r \eta_m \eta_R \eta_D}{R D}$

$\therefore \dfrac{P_{up}}{P_{down}} = \dfrac{1}{(\eta_m \eta_R \eta_D)^2} = \dfrac{1}{\eta_T^2}$

4.3 Numerical Values

Data

$r = 0.25$ m
$M = 2.5$ Tonne
$D = 574$ cm^3 rev^{-1} = 9.1 x 10^{-5} m^3 rad^{-1}
$R = 5:1$ (reduction)
$\eta_m = 0.78$ (starting), 0.92 (running)
$\eta_R = \eta_d = 0.94$

Performance Analysis

Ideal pressure

$$P = \frac{M g r}{R D} = \frac{2500 \times 9.81 \times 0.25}{5 \times 9.1 \times 10^{-5}} = 133 \text{ bar}$$

Therefore, during start-up from rest:

$\eta_T = 0.69$
$P_{up} = 193$ bar & $P_{down} = 92$ bar

And when operating at speed:

$\eta_T = 0.81$
$P_{up} = 164$ bar & $P_{down} = 108$ bar

5. Hydraulic Motor for Driving a Winch

A winch is to be driven by a hydraulic motor and in order to provide a wide range of operating speeds at the maximum supply flow the motor displacement can be set at either a maximum or minimum value. These displacement values (D_{max} and D_{min}) are pre-set by mechanical stops in the motor but their level can be chosen from within the range shown in the data. The winch operator makes the selection of the two displacements by changing the position of a control valve in the hydraulic circuit.

The mechanical transmission has a speed-reducing gearbox and to design the system it is required to select an appropriate motor from the two sizes given in the data and determine the reduction ratio of the gearbox that will provide the specified performance.

1) Calculate the gearbox ratio that is required for each motor to give the maximum winch torque when the motor is operating at a selected pressure.

2) Determine the motor speeds from 1) that are required to give the maximum cable speed of 50m/min and select a suitable motor from the data.

3) Calculate the flow that is required to drive the selected motor when in maximum displacement at the speed that produces a cable speed of 15m/min.

4) Using the flow from 3) determine the required minimum motor displacement that will provide the maximum cable speed of 50m/min.

5) For this flow, calculate the motor speed for an oil viscosity of 10cSt. This needs to account for the effect of oil viscosity on the volumetric efficiency.

Data

Motors (Maximum pressure for continuous operation = 300bar)

Motor displacement (D_{max}/D_{min}) cm³/rev	55/11	80/17
Maximum speed at D_{max} rev/min	4200	3750
Maximum speed for $D < D_{max}$ rev/min	6300	5600

Winch drum diameter 0.5m
Gearbox mechanical efficiency (η_g) 92%
Maximum winch load 200kN
Maximum cable speed at maximum load 15m/min
Maximum cable speed 50m/min
Motor mechanical efficiency at a) D_{max}, b) $D<D_{max}$ (η_H) 95%, 90%
Motor volumetric efficiency for D_{max} (35 cSt oil viscosity)(η_{VH}) 95%
Motor volumetric efficiency for $D < D_{max}$ (35 cSt oil viscosity)(η_{VL}) 90%

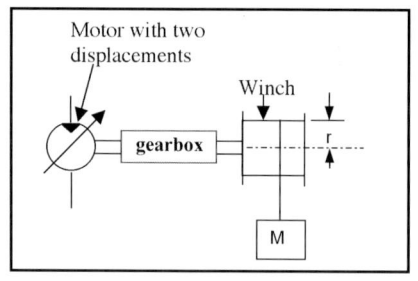

5.1 Gearbox ratio

Choose the maximum operating pressure of 300bar.
Winch drum torque = Load force x drum radius = $200 \times 10^3 \times \dfrac{0.5}{2} = 50 \times 10^3$ Nm
Motor torque at 300bar:

$$D_{max} \times \frac{10^{-6}}{2\pi} \times 300 \times 10^5 \times \eta_H(0.95) \times \eta_G(0.92) = 4.17 D_{max} \text{ Nm}$$

Required reduction gearbox ratio (n) $\dfrac{\text{winch torque}}{\text{motor torque}} = \dfrac{T_D}{T_m} = \dfrac{50 \times 10^3}{4.17 D_{max}}$

This gives values of n required for the two types of motor of 218 and 150.

Performance Analysis

5.2 Motor selection

Cable speed $\pi dN = \dfrac{\pi \times 0.5 N_m}{50n} = U_{max} = 50m/min$ (N_m = motor speed)

Motor speed $N_m = \dfrac{0.5\pi}{}$ which gives motor speeds of 6939 and 4775 rev/min respectively

Select the larger motor, as its speed is less than the maximum allowable value given in the data.

5.3 Flow required

Motor speed for a cable speed of 15m/min $\dfrac{15}{0.5\pi} \times 150 = 1432 \, rev/min$

The flow, Q, required for this speed $= \dfrac{N_m D_{max}}{\eta_{VH}} = \dfrac{1432 \times 80 \times 10^{-3}}{0.95} = 121 L/min$

5.4 Minimum motor displacement

For the flow given in 5.3, the minimum motor displacement required to operate the winch at a cable speed of 50 m/min is given by:

$$D_{min} = \dfrac{Q}{N_m} \times \eta_{VL} = \dfrac{121 \times 10^3 \times 0.9}{4775} = 23 cm^3/rev$$

5.5 Maximum motor speed with oil having a viscosity of 10cSt

At minimum displacement the volumetric efficiency of 90% was quoted for an oil viscosity of 35cSt which represents a leakage flow of 10%. This leakage will increase as the oil viscosity reduces so that here it will be $= 10\% \times \dfrac{35}{10} = 35\%$ and the volumetric efficiency will therefore be 65%.

6. Hydraulic system for gantry crane

A hydraulically powered gantry crane, shown in Figure 2, is to lift loads up to 300 Tons. It is required to design the hydraulic systems for driving both the winch and the crane wheels.

Details of the installation are given below:

184   Principles of Hydraulic System Design

Figure 2. Gantry crane

a) Crane Winch (2 Drums)

- Total load   = 300 Tonne

Two pulley blocks each having 4 pulleys to provide 8 lengths of cable for each of the two lifting hooks.

- Maximum lift   = 10m
- Winch drum diameter   = 600mm
- Lift speed (variable) (max) = 0.02m/s

Each of the two winch drums are to be powered separately, and are required to be held stationary in any position.

b) Wheel Drive

- The vehicle has eight wheels that are grouped together into units having two wheels each. In these units the two wheels are connected by a chain drive, with one of the wheels being driven by a hydraulic motor (i.e. four motors in total). There is no requirement for brakes on these wheels.
- Supply pipe lengths are 30m on one side and 80m on the other.

Performance Analysis

- As far as is possible the torque on each side must be within 10% of each other.
- Braking will be carried out by reducing the pump supply and means must be provided to prevent both cavitation and excessive pressures.
- Gear reduction ratio between wheels and motor = 2.5:1
- Wheel drive system efficiency = 80%
- Maximum wheel torque at low speed = 4250Nm
- Maximum wheel torque at high speed = 2120Nm
- Wheel diameter = 0.46m
- Maximum speed = 30m/min
- Minimum speed = 1.2m/min
- Maximum pressure = 250bar

Cable diameter *mm*	Minimum breaking force (*kN*)
35	785
40	1000
48	1460

Cable Information

6.1 Gantry crane

Load = *300 tonne* = 3000kN

8 Cables (falls) and 2 pulley blocks

Cable tension = $\frac{3000}{2 \times 8}$ = *187.5kN*

Allow 10% for friction - *206.3kN*

Choose cable of 40mm diameter

For one cable layer the winch drum torque = $\left(\frac{0.6 + 0.04}{2}\right) \times 206.3 \times 10^3$ = *66,000Nm*

The winches can be driven by either low or high speed motors and the reduction gearbox ratio needs to be selected accordingly.

a) Low speed motor

Choose 10:1 reduction gearbox and allow 8% for mechanical losses

The motor torque $= \dfrac{66000}{10} \times \dfrac{1}{0.92} = 7{,}176 Nm$

For 80% motor efficiency at 250bar the motor displacement is given by:

$$Motor\ torque\ T = PD\eta_m$$

$$\therefore D = \dfrac{7176}{250 \times 10^5 \times 0.8} = 3.6 \times 10^{-4}\ m^3/rad$$

$$= 2254(cm^3/rev)\ or\ 2.254(L/rev)$$

There are 4 pulleys and 8 lengths of cable on each side which gives a maximum cable speed of:

$$0.02 \times 8 = 0.16 m/s$$

$\therefore$ The winch drum speed $= \dfrac{0.16 \times 60}{\pi \times 0.64} = 4.78 rev/min$ and the motor speed will be 47.8 rev/min.

b) High-speed motor

For a high-speed motor, a gearbox ratio of 100:1 can be used which, using the same gear box efficiency as in a), will require a motor having a displacement of 225 cm³/rev operating at a speed of 478 rev/min.

c) Motor flow

For both the high and low speed motor drives, the motor flows will be the same. Thus, assuming a motor volumetric efficiency of 90% the flow required for each winch is:

$$\dfrac{speed \times displacement}{volumetric\ efficiency} = \dfrac{478 \times 0.225}{0.9} = 119.5 L/min$$

The required maximum pump displacement for supplying both winches, assuming a volumetric efficiency of 95% and a drive speed of 1500 rev/min, will be:

$$\dfrac{119.5 \times 10^3}{1500 \times 0.95} \times 2 = 167.7\ cm^3/rev$$

Performance Analysis

For a mechanical efficiency of 95% for the pump the required input power is:

$$= \frac{2 \times 119.5 \times 250 \times 10^5}{60 \times 10^3 \times 10^3 \times 0.95} = 104.8 kw$$

The total output power from the winches is:

$$3 \times 10^6 \times 0.02 \times 10^{-3}\ kw = 60\ kw$$

Appropriate winch motors and supply pump can be selected from manufacturers' literature and the assumed values of the mechanical and volumetric efficiencies used in the analysis can be checked.

6.2 Wheel drive

For a torque at each wheel of 4250Nm, two wheels require 8500 Nm torque. Thus for a gearbox reduction ratio of 2.5:1 the required motor torque is given by:

$$\frac{8500}{2.5 \times 0.8} = 4250Nm \quad (80\% \text{ wheel drive mechanical efficiency})$$

At a vehicle speed of 1.2m/min (U) with 0.46m wheel diameter, $\pi DN = U$

Hence $N = \frac{U}{\pi D} = \frac{1.2}{\pi \times 0.46} = 0.83 rev/min$

U(m/min)	1.2	30.0
Wheel Speed (rev/min)	0.83	20.75
Motor Speed (rev/min)	2.08	51.9
Motor D_m (cm³/rev)	1476	738
Nominal Flow (L/min)	3.06	38.3
Motor Torque (Nm)	4250	2120
Pressure (bar)	215	214

Table 1. Motor performance

As for the winch drive, low or high-speed motors can be selected for the wheel drive. By way of example, a dual displacement low speed motor is considered here which gives the performance outlined in Table 1.

The maximum total nominal flow (two motors each side) = 2 x 38.3 x 2 = 153.2L / min. Assuming that the volumetric efficiency of the motors is 90% and 95% for the pump gives a maximum pump flow of:

$$\frac{153.2}{0.9 \times 0.95} = 179 L/min$$

For 1500 rev/min pump speed, the required pump displacement is:

$$D_p = \frac{179 \times 10^3}{1500} = 119.5 cm^3/rev$$

As for the winch drive, the pump and motors having displacements that are closest to those required can be selected from manufacturers' literature.

6.3 Pipe sizes

For motor flows of 80L/min to each side.

The pressure loss $\Delta_p = 4f \frac{L}{d} \frac{\rho}{2} U^2$

f - Friction factor from Moody diagram (chapter 8)

For the pipe the value of the Reynolds No. $R_e = \frac{Ud}{\upsilon}$

Taking a value of fluid viscosity of 40cSt ($4 \times 10\text{-}5 m^2/s$) and length L of the longest side of 80m gives:

$$Q = \frac{\pi d^2}{4} U \qquad \therefore \Delta p = 8 \times (4f) \frac{L}{d^5} \frac{\rho}{\pi^2} Q^2$$

d mm	Re	4f	Δ_p bar	U m/s
20	2122	0.03	9.4	4.2
25	1700	0.038	3.9	2.7
30	1415	0.043	1.8	1.9

Table 2 Pipe pressure loss

The effect of the pipe diameter on the pressure loss can be seen from Table 2.

Performance Analysis

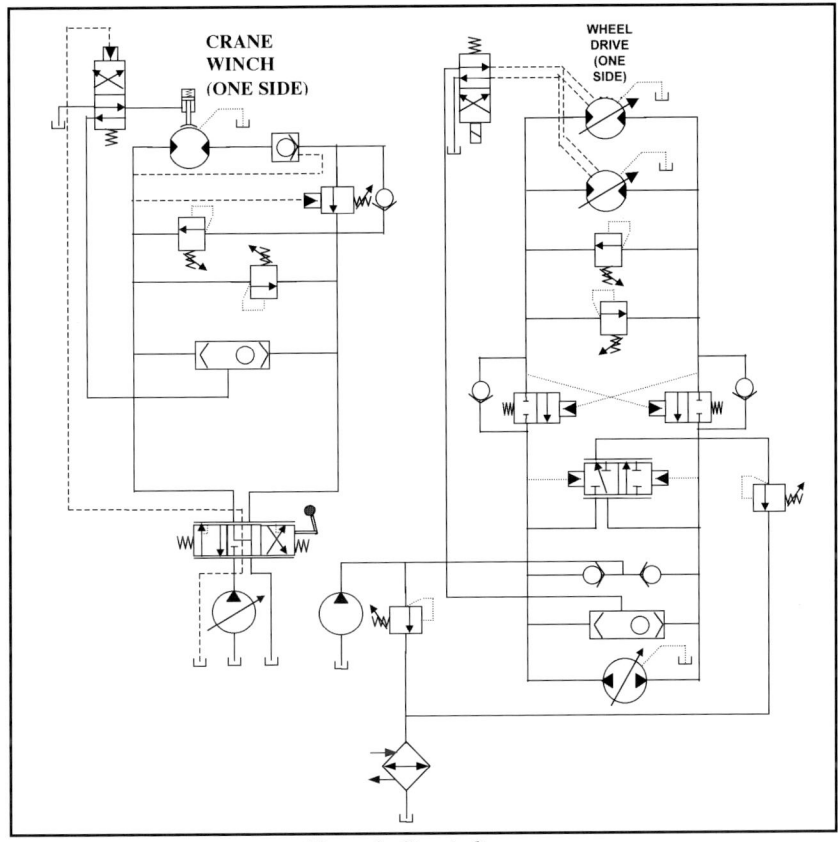

Figure 3. Circuit diagram

A circuit that will provide the necessary functions is shown in Figure 3 This contains the following major functions:

Crane Winch

- Pilot operated check valve to protect from hose failures
- Cross line relief valves
- Brake control by:
    - Selection of the DCV
    - High pressure in the circuit
- Counterbalance valves for lowering the load

Wheel drive

- Cross line relief valves
- Brake valves
- Purge valve to extract flow for cooling from the low pressure side of the circuit (drain flows may also be passed to the cooler inlet).
- Motor displacement selection valve.
- Boost pump to make up drain flows from the pump and motors.

7. Pressure losses

The evaluation of the pressure losses in pipes was described in Chapter 8 and a typical problem area for the application of this analysis is the determination of appropriate pipe sizes for the inlet, or suction, flow to pumps directly from a reservoir. Frequently there are valves and pipe fittings in such circuits and standard charts can be used to give the dynamic head losses that will arise. This example is concerned with the evaluation of a typical pump application.

The connections to a pump inlet from the supply tank consist of the following components:

- A pipe, descending vertically having a length of 1m and an internal diameter of 25mm
- A gate valve which is fully opened
- A standard sweep elbow having an internal diameter of 25mm
- A horizontal pipe having a length of 3m and an internal diameter of 25mm

The supply tank is not pressurised.

The pump is required to operate with a fluid temperature range from 20^0C to 60^0C. Using the given data and the graphs in chapter 8 calculate the pressure at the pump inlet for fluid temperatures of 20^0C and 60^0C. Pipe inlet effects can be ignored.

The pump requires an inlet pressure of at least 0.9 bar absolute (-0.1 bar gauge) to operate satisfactorily. Show how the system can be modified to achieve this.

7.1 Data

Pump displacement   48 cm^3/rev
Pump speed   1500 rev/min

7.2 Pressure loss at 20⁰C

The pump flow $Q = 48 \times 10^{-6} \times \dfrac{1500}{60} = 1.2 \times 10^{-3} \, m^3/s$

Pipe area (d = 25mm) $A = \dfrac{\pi \times 25^2 \times 10^{-6}}{4} = 4.91 \times 10^{-4} \, m^2$

The mean velocity in the pipe $u = \dfrac{1.2 \times 10^{-3}}{4.91 \times 10^{-4}} = 2.44 \, m/s$

At 20⁰C the fluid viscosity from Figure 1, Chapter 8 is 85 cSt.

The Reynolds No. $R_e = \dfrac{ud}{\upsilon} = \dfrac{2.44 \times 0.025}{85 \times 10^{-6}} = 717$

Therefore the flow is laminar for which the friction coefficient $f = \dfrac{16}{R_e} = 0.022$

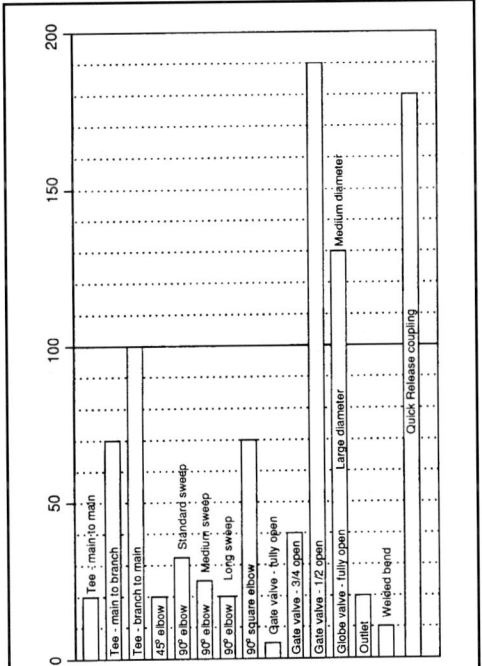

Figure 4. Fluid Resistance in Fittings

From the chart in Figure 4 the fluid resistance of fittings is given as an equivalent length of pipe for which the pressure loss is calculated in the normal manner for a pipe. For the circuit details given in the data we get the following equivalent pipe lengths for the various fittings from Figure 4:

- 5 pipe diameters for the fully opened gate valve
- 32 pipe diameters for the standard sweep elbow

The equivalent length, L, of the fittings is therefore $(5 + 32)d$ giving an $\frac{L}{d}$ ratio of 37 which is added to that for the pipe system.

The total pressure loss:

$$= 4f \frac{L}{d} \frac{\rho u^2}{2} = 4 \times 0.022 \times \left(\frac{4}{0.025} + 37\right) \frac{875 \times 2.44^2}{2} = 44900 \, pa \; (= 5.1 \times 4f \, bar)$$

The pressure increase from the vertical height of the supply tank $= \rho g h = 870 \times 9.81 \times 1 = 8580 \, pa$. The pump inlet pressure $= 8580 - 44900 = -0.36$ bar (gauge).

7.3 Pressure loss at 60°C

At a fluid temperature of 60°C the viscosity is 15 cSt for which the Reynolds No. is $717 \times \frac{85}{15} = 4100$. Therefore, the flow is turbulent, for which value of the friction factor f obtained from the graph in Figure 2, chapter 8 is approximately 0.01. For this condition the friction pressure loss is:

$$5.1 \times 0.04 = 0.204 \, bar$$

The inlet pressure will therefore be: $-0.204 + 0.086 = -0.118$ bar (gauge)

7.4 Pump requirements

Both of the calculated inlet pressures are lower than those recommended and in order to reduce the pressure loss the pipe diameter can be increased. Increasing the pipe diameter to 35mm will reduce the pressure loss for the 20°C condition to the value given as follows:

$$\text{New velocity} = 2.44 \times \left(\frac{25}{35}\right)^2 = 1.24 \, m/s$$

Performance Analysis

This gives a new Reynolds Number $R_e = \dfrac{ud}{v} = \dfrac{1.24 \times 0.035}{85 \times 10^{-6}} = 510$

$\therefore f = \dfrac{16}{510} = 0.031$ for which the pressure loss will be:

$$4f\dfrac{L}{d}\dfrac{\rho u^2}{2} = 4 \times 0.031 \times \left(\dfrac{4}{0.035} + 37\right)\dfrac{875 \times 1.24^2}{2} = 12620\, pa$$

The inlet pressure will now be 0.086 - 0.126= - 0.04bar (gauge) which is less than the minimum allowable and therefore satisfactory.

8. Single stage relief valve

For a single stage relief valve, having the given data, calculate the variation of the flow through the valve and the valve opening with pressure. The spring preload is set to give a cracking pressure of *200 bar*.

Data:

Valve seat diameter	*12.5 mm*
Valve poppet seat angle	*45⁰*
Spring stiffness	*100N/mm*
Valve flow coefficient	*0.8*

For a direct acting, poppet type of relief valve the equation for the variation in flow with pressure including the effect of the flow force is, from chapter 8, 6:

$$Q = \dfrac{P - C/A}{K_1/\sqrt{P} + K_2\sqrt{P}}$$

For the data the values of the constants are:

$$K_1 = \dfrac{k}{A\pi dC_Q \sin\phi \sqrt{\dfrac{2}{\rho}}} = 7.7 \times 10^{11}$$

$$K_2 = \frac{\cos\phi}{A}\sqrt{2\rho} = 2.4 \times 10^5$$

$$X = \frac{Q}{\pi d C_Q \sqrt{\frac{2P}{\rho}} \sin\phi}$$

P bar	Q m³/s	Q L/min	X mm
210	7.9x10⁻⁴	47	0.17
220	1.55x10⁻³	93	0.31
250	3.7x10⁻³	222	0.70

9. Simple actuator cushion

Using the analysis given in Chapter 3, 4.2, the performance of an actuator cushion using a simple orifice or adjustable restrictor valve can be determined for the actuator application having the given data. The peak actuator pressure needs to be limited to 300 bar that occurs at the commencement of cushioning.

Data:

Actuator piston diameter   *140 mm*
Cushion plug diameter   *43 mm*
Mass   *40 000 kg*
Initial velocity   *0.3 ms⁻¹*

Performance Analysis

From Chapter 3, 4.2, the velocity variation is given by:

$$U = U_m \exp(-\frac{CX}{m})$$

where $C = \dfrac{\rho A_C^3}{2 C_D^2 A_R^2}$

For a maximum pressure of *300 bar* the restrictor size is given by:

$$A_R = \frac{Q_m}{C_D}\sqrt{\frac{\rho}{2 P_C}}$$

$$A_C = \frac{\pi}{4}(d_P^2 - d_C^2) = 13.9 \times 10^{-3} \, m^2$$

$$Q_m = A_C U_m = 0.3 \times 13.9 \times 10^{-3} = 4.2 \times 10^{-3} \, m^3 s^{-1} \quad (251 \, L/\min)$$

and, taking a value for C_D of 0.65 gives:

$$A_R = \frac{4.2 \times 10^{-3}}{0.65}\sqrt{\frac{870}{2 \times 3 \times 10^7}} = 24.6 \times 10^{-6} \, m^2 \quad (5.6 \, mm \, diam. \, hole)$$

Now:
$$U = U_m \exp(-\frac{CX}{m})$$

where $C = \dfrac{\rho A_C^3}{2 C_D^2 A_R^2} = 4.6 \times 10^6$

$$\frac{m}{C} = \frac{40000}{4.6 \times 10^6} \times 10^3 \, mm = 8.6 \, mm$$

$$\therefore U = 0.3 \exp\left[-\frac{x}{8.6}\right] \, m/s$$

The velocity will reduce to 37% of its initial value (*0.11 m/s*) in *8.6 mm*.

10. Central bypass valve

Figure 5 shows a weight-loaded actuator that is operated by a central bypass type of the open centre valve.

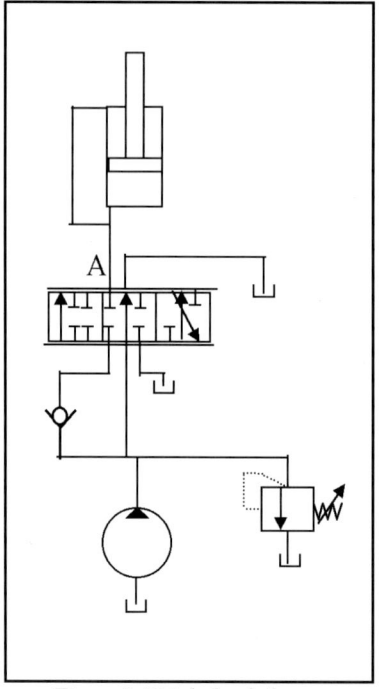

Figure 5. Weight loaded system

The flow characteristics for central bypass valves are usually given in manufacturers' literature. However, an analysis of these valves provide a good example of the use of variable restrictions for the control of flow and how the valve characteristics are obtained.

Figure 6 shows the basic valve dimensions where the distance L refers to the initial opening of the bypass and the distance H refers the overlap of the metering to the outlet port.

The system circuit can be reduced to that shown in Figure 7 where the variable restrictions are the bypass and outlet port metering openings that are altered by movement of the valve spool, that of the bypass reducing as the spool is progressively moved from the central position.

Performance Analysis

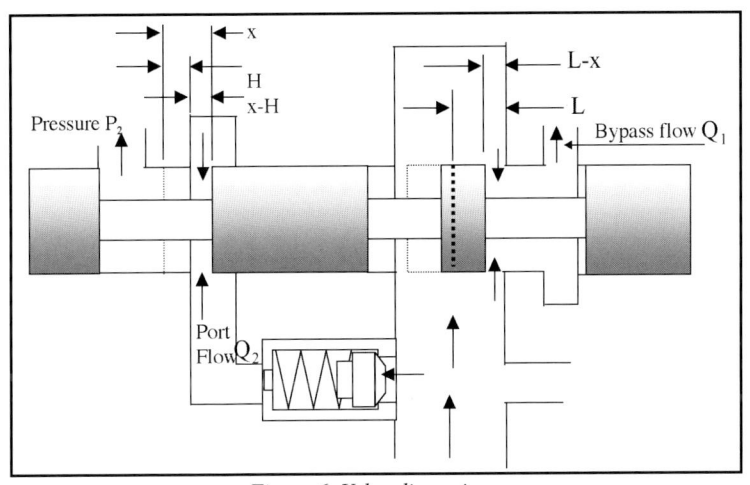

Figure 6. Valve dimensions

As the valve is moved through distance H the pump pressure is increased because of the reduced bypass opening. For movements greater than H, flow can pass to the outlet port providing, as discussed in Chapter 5, the pump pressure can open the load check valve against the pressure required to operate the actuator.

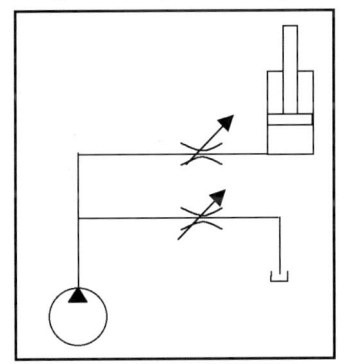

Figure 7. Equivalent circuit for actuator extension

10.1 Flow analysis

Assuming that the outlet pressure from the central bypass is small (i.e. zero pressure losses to the tank) we get:

Bypass flow $\quad Q_1 = K(L - x)\sqrt{P_s}$ \quad (1)

Port flow $$Q_2 = K(x-H)\sqrt{(P_S - P_2)} \qquad (2)$$

Pump flow $$Q_P = Q_1 + Q_2 \qquad (3)$$

Note that for $x < H$, Q_2 is zero because the valve is positioned in the overlap region. The flow constant $K = C_Q \pi d \sqrt{\dfrac{2}{\rho}}$

We can obtain solutions to these equations for a given valve position and load pressure, P_2, which would give the pump pressure and hence the values of Q_1 and Q_2. However, carrying out this process is complicated because of the square root of pressure. An alternative method can be applied that uses two known extreme conditions to establish the range of values of Q_2 for any given valve opening, x.

Condition 1

Condition 1 establishes the pump pressure that is created when all the pump flow is passed through the bypass. Thus from equation 1 we have:

$$P_S = \left[\frac{Q_1}{K(L-x)}\right]^2 = \left[\frac{Q_P}{K(L-x)}\right]^2 \qquad (4)$$

Equation 4 determines the maximum pump pressure that is available for a given value of x, which establishes the valve movement that is required in order to obtain flow at the valve outlet for a given load pressure, P_2. On further movement of the valve beyond this point, flow will pass to the outlet.

Condition 2

The maximum available outlet flow, Q_{2max}, will occur with zero load pressure and this condition can be obtained from equations 1 to 3 with $P_2 = 0$.

Thus, equations 1 and 2 give:

$$P_S = \left[\frac{Q_{2max}}{K(x-H)}\right]^2 = \left[\frac{Q_1}{K(L-x)}\right]^2$$

Performance Analysis

Or: $\quad Q_{2max} = \left(\dfrac{x-H}{L-x}\right) Q_1\quad$ which into equation 3 gives:

$$Q_{2max} = \left(\dfrac{x-H}{L-H}\right) Q_P \tag{5}$$

Equation 5 shows that the maximum possible outlet flow varies linearly with x, for zero outlet pressure, when x > H (i.e. the valve has moved out of the overlap region) reaching its maximum value (pump flow) when the central bypass is fully closed (i.e. x = L).

Practical example

For the system in Figure 5 the valve and actuator dimensional data are given as follows:

Data

Pump flow	= 400 L/min
Valve spool diameter	= 12 mm
Central bypass opening for valve in central position	= 6 mm
Valve overlap at port A	= 2 mm
Actuator rod diameter	= 70 mm
Actuator piston diameter	= 100 mm

10.2 Valve characteristics

Consider the extension of the actuator for lifting the weight load. We can apply conditions 1 and 2 in order to establish the operating flow range of the valve at different positions. Thus we have:

$$K = C_Q \pi d \sqrt{\dfrac{2}{\rho}} = 0.65 \times \pi \times 12 \times 10^{-3} \sqrt{\dfrac{2}{870}} = 1.18 \times 10^{-3} m^3 / s /(N/m^2) \tag{6}$$

For condition 2 from equation 5 it is seen that, for zero load pressure, the outlet flow, Q_2, varies linearly with valve movements from x = 2mm to x = 6mm.

For condition 1 the variation of pump pressure for zero outlet flow is obtained from equation 4. Thus:

$$P_S = \left(\frac{Q_P}{K(L-x)}\right)^2 = \left(\frac{400}{1.18 \times 10^{-3} \times 6 \times 10^4 \times (6-x) \times 10^{-3}}\right)^2 \times 10^{-5} \, bar = \frac{319}{(6-x)^2} \, bar \quad (7)$$

Values from equation 7 are shown in Table 1 where it can be seen that the pump pressure increases rapidly as x nears to the closing position of the bypass. If the valve is moved rapidly to the full open position (bypass fully closed), excessive transient pump pressures could occur. In order to avoid this and for when there is a blocked outlet port or an excessive actuator load force, a relief valve needs to be fitted.

Valve position x (mm)	Maximum flow (for $P_2=0$) Q_{2max} L/min	Pump pressure (for $Q_1=Q_p$) P_p (bar)
0	0	8.9
1	0	12.8
2	0	19.9
3	100	35.4
4	200	80.0
5	300	319.0
6	400	∞

Table 1

10.3 Actuator force

The actuator has a piston diameter of 100mm and a rod diameter of 70mm and for the connections shown in Figure 5 (regenerative) the effective area is that of the rod. This is because the supply pressure acts on both the piston and annulus areas and the net flow into the actuator is due to the volume displaced by the rod area only.

The rod area $= \frac{\pi}{4} \times (100^2 - 70^2) \times 10^{-6} = 4 \times 10^{-3} \, m^2$

The actuator weight load capacity can be determined at a given valve position from Table 1, the value chosen being dependent on the maximum required velocity.

Performance Analysis

Thus for a load pressure of 35.4 bar (corresponding to the maximum possible at x = 3 mm), the weight load that can be lifted is:

$$F = 4 \times 10^{-3} \times 35.4 \times 10^5 = 14160\,N$$

10.4 Valve operation

The valve will need to be opened by 3 mm in order to generate the necessary pump pressure of 35.4 bar. The flow that is available at valve positions greater than 3mm can be determined by the following method.

x = 4mm

L - H = 4mm and x – H = 2mm.

Thus, from equation 5, Q_{2max} = 200 L/min (valve at mid open position).

From Table 1, the maximum pump pressure for x = 4mm is 80bar and the pressure for other bypass flows can be obtained from equation 7 thus:

$$P_S = 80 \times (\frac{Q_1}{400})^2\,bar \qquad (8)$$

And from equation 2,

$$P_S - P_2 = \left(\frac{Q_2}{K(x-H)}\right)^2 \qquad (9)$$

Thus by selecting values of Q_2 between 0 and 200 L/min values of P_S and P_2 can be obtained from equations 8 and 9 and these are shown in Table 2.

Q_2 L/min	Q_1 L/min	P_S bar	$P_S - P_2$ bar	P_2 bar
200	200	20	20	0
150	250	31.3	11.3	20
100	300	45	5	40
50	350	61.3	1.3	60
0	400	80	0	80

Table 2

For x = 5mm
Q_{2max} = 300 L/min
 Applying the same method as for x = 4mm gives the values shown in Table 3.

Q_2 L/min	Q_1 L/min	P_S bar	$P_S - P_2$ bar	P_2 bar
300	100	20	20	0
200	200	80	8.9	71.1
100	300	180	2.2	177.8
0	400	319	0	319

Table 3.

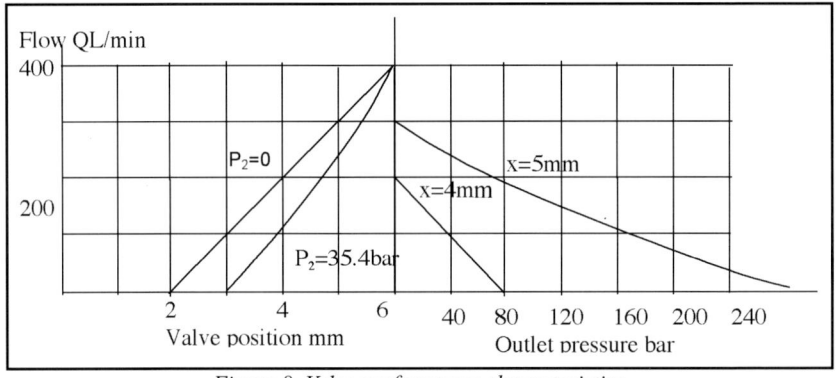

Figure 8. Valve performance characteristics

From Figure 8 it is seen that for the weight load of 14160 N, the change of flow with valve position is reasonably linear for the 3mm of valve movement required for zero to maximum flow. If a heavier weight load is lifted, say that for 80bar pressure, the valve will have to be opened 4mm before the start of actuator movement.

For lowering the load, the control of flow is made by metering the outlet flow to the return, the valve spool being arranged so that pump flow is bypassed to the return for this position.

11. Pump and motor efficiencies

a)

The pistons of an oil-hydraulic pump are 20 mm diameter and 25 mm effective length. At the rated performance, the mean piston velocity is 1 m/s and the pump

Performance Analysis

discharge pressure is 200 bar. Estimate the radial clearance between the piston and the cylinder bore that will give optimum overall efficiency. For this application the fluid kinematic viscosity, ν, is 35 cSt and the fluid kinematic density is 870 kg/m³. The dynamic viscosity $\mu = \nu \rho = 35 \times 10^{-6} \times 870 = 0.03 Nsm^{-2}$.

For a leakage flow of Q_s with a pump pressure of P the leakage power loss is given by:

$$Q_s P = (\frac{h^3}{12\mu} \times \frac{P}{x} \times 2\pi R) \times \frac{P}{2} \text{ (piston pressurised for half the time)}.$$

$$\frac{h^3}{12 \times 0.03} \times \frac{2 \times 10^7}{25 \times 10^{-3}} \times 2\pi \times \frac{20}{2} \times 10^{-3} \times \frac{2 \times 10^7}{2} = 1.4 \times 10^{15} h^3 \text{ W}.$$

The frictional power loss is caused by the relative velocity between the piston and the cylinder in the clearance space acting on the surface areas. This creates a shear stress in the fluid that acts over the surface area. The viscous shear stress, τ, in the fluid is given by:

$$\tau = \mu \frac{\partial u}{\partial y} \qquad \text{Couette flow} \quad \tau = \mu \frac{U}{h}$$

Thus this power loss is given by:

$$\tau AU = (\mu \frac{U}{h})(2\pi RL) U$$

$$= \frac{1}{h} \times 0.03 \times 1 \times 2\pi \times \frac{20 \times 10^{-3}}{2} \times 25 \times 10^{-3} = 4.71 \times 10^{-5} \times \frac{1}{h} \text{ W}$$

(for $U = 1 m/s$)

Adding the viscous and leakage power losses and differentiating with respect to h will give a minimum when this is zero. Thus we have:

$$3 \times 1.4 \times 10^{15} h^2 - 4.71 \times 10^{-5} \times \frac{1}{h^2} = 0$$

$$h^4 = 1.12 \times 10^{-20} \quad h = 10.3 \mu m$$

b)

A pump which runs at 1450 rev/min has the following loss coefficients:
 Slip coefficient $C_s = 1.2 \times 10^{-8}$ Coulomb friction coefficient $C_f = 0.03$
 Viscous friction coefficient $C_v = 1.31 \times 10^5$
 Oil kinematic viscosity = 35cSt
 Estimate the maximum overall efficiency of the pump and the corresponding discharge pressure.

For maximum efficiency, $C_S\left(\dfrac{P}{\mu\omega}\right) = C_V\left(\dfrac{\mu\omega}{P}\right)$

$$P^2 = \dfrac{C_V}{C_S}(\mu\omega)^2 = \dfrac{1.3 \times 10^5}{1.2 \times 10^{-8}}\left[\dfrac{0.03 \times 1450 \times 2\pi}{60}\right]^2$$

$$P^2 = 2.25 \times 10^{14} \quad P = 150\,bar$$

Slip loss = Viscous friction loss = $C_S\left(\dfrac{\mu\omega}{P}\right) = 0.04$

Overall efficiency = $\dfrac{1 - 0.04}{1 + 0.03 + 0.04} = 0.9$

12. Control System Design

From Chapter 10, equation 38, a valve/actuator system has an open-loop system frequency response relating the actuator position to the valve spool position as follows:

$$R = \left|\dfrac{y}{x}\right| = \dfrac{k}{\omega\sqrt{\left[1 - \left(\dfrac{\omega}{\omega_n}\right)^2\right]^2 + \left[\dfrac{2\zeta\omega}{\omega_n}\right]^2}}$$

12.1 Data

Actuator

Supply pressure 210 bar.
Actuator: Piston area $10^{-3}\ m^2$
Total volume (V_T) of trapped oil 0.3 L
For the actuator piston in the centre the volume (V) each side is 0.15 L
Fluid bulk modulus $1.4 \times 10^9\ N/M^2$

Valve

Flow coefficient C_x $0.96\ m^2/s$
Flow coefficient C_p $7.42 \times 10^{-12} (m^3/s)/(N/m^2)$
Load:
Mass 250 kg
Viscous friction 16 kN/(m/s)

Performance Analysis

Natural frequency

$$\omega_n = \left[\frac{2A^2B}{MV}\right]^{1/2} \text{ or } \left[\frac{4A^2B}{MV_T}\right]^{1/2}$$

$$\omega_n = \left(\frac{2 \times (10^{-3})^2 \times 1.4 \times 10^9}{250 \times 0{,}15 \times 10^{-3}}\right)^{0.5} = 273 \, rad/s$$

Damping factor

$$\frac{2\zeta}{\omega_n} = \frac{C_f V}{2\beta A^2} + \frac{MC_P}{2A^2}$$

$$\zeta = \left(\frac{16 \times 10^3 \times 0.15 \times 10^{-3}}{1.4 \times 10^9} + (7.42 \times 10^{-12} \times 250)\right)\left(\frac{273}{2 \times 2 \times 10^{-6}}\right) = 0.243$$

12.2 System gain

The gain, k, is $= \dfrac{C_x}{A} = \dfrac{0.96}{(1 \times 10^{-2})} = 960 \, s^{-1}$

Hence the amplitude ratio and phase angle can be found by substituting in different values of ω into the following equations:

$$R = \frac{960}{\omega\sqrt{\left[1-\left(\frac{\omega}{273}\right)^2\right]^2 + \left[\frac{0.486\,\omega}{273}\right]^2}} \quad \text{and} \quad \phi = -90 - \tan^{-1}\left[\frac{\frac{0.486\,\omega}{273}}{1-\left(\frac{\omega}{273}\right)^2}\right]$$

The resulting Bode plot in Figure 9 shows that the system would be unstable if it were operated in a closed-loop. Therefore, the system must be stabilised by reducing the gain, *k* (In practice, this could be achieved by reducing the supply pressure or the valve flow gain).

Thus, the amplitude ratio when $\phi = -180$ (i.e. $\omega = \omega_n$) is 17.2 dB. Therefore, to provide sufficient stability margin, k has to be reduced by:

$$17.2 + 6 = 23.2 \, dB$$

$$\therefore 20 \log 960 = 20 \log k - 23.2$$

$$\frac{23.2}{20} = 1.16 = \log \frac{960}{k}$$

$$\frac{960}{k} = 10^{1.16} = 14.45$$

Hence, the value of k to give an adequate stability margin must be:

$$k = \frac{960}{14.45} = 66.5 \, rad \, s^{-1}$$

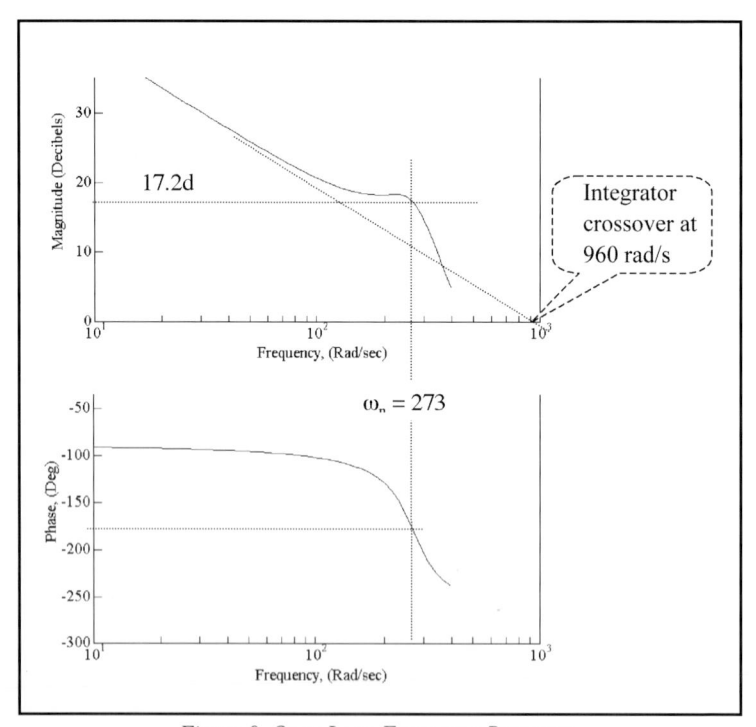

Figure 9. Open Loop Frequency Response

In the system as represented in Figure 9, there will be a power amplifier to deliver current to the valve. This amplifier will have an adjustment facility that will allow the gain to be set at the required value.

13. Hydraulic system for injection moulding machine

A hydraulic system is required to operate an injection-moulding machine that has a movable platen using two hydraulic cylinders for the mould and an injection screw feed of granular material through the heater. Design data is given below:

13.1 System data

- Maximum operating pressure = 150 bar
- Platen clamping force (to be pre-set) = 150000 N(max)
- Platen movement, l = 50mm
- Time required for platen movement (t_c) = 0.1s
- Clamping time = 1.2s
- Time for injection of material (t_I) = 0.1s
- Feed screw torque = 1500Nm
- Feed screw operating speed = 200rev/min
- Mass of platens = 250kg
- Injection screw operates after the platens have closed

13.2 Injection moulding machine

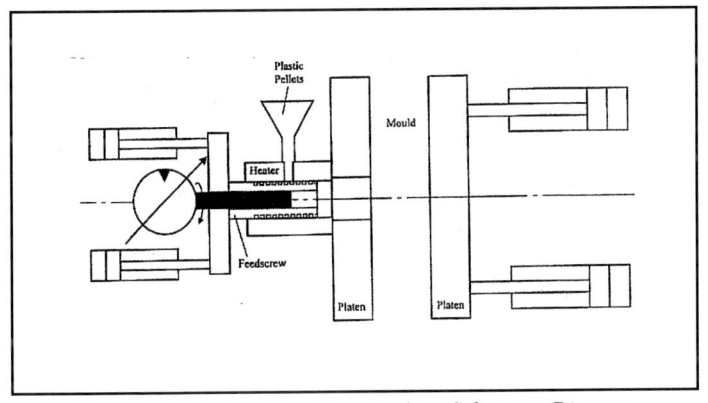

Figure 10. Injection Moulding Machine Schematic Diagram.

13.3 Actuator

Required clamping force = 150,000N

Total actuator area, $A = \dfrac{150000}{150 \times 10^5}$ for 150 bar supply pressure $\therefore A = 10^{-2} m^2$

The flow required to move two cylinders,

$$Q_A = \left(\dfrac{Al}{t_c}\right) = \dfrac{10^{-2} \times 50 \times 10^{-3} \times 60 \times 10^3}{0.1}$$

$$Q_A = 300 \; l/min$$

13.4 Motor

The motor displacement required, $D_m = \dfrac{T_m}{P} = \dfrac{1500 \times 2\pi \times 10^3}{150 \times 10^5 \times 0.8} = 0.785 \; L/rev$ for a mechanical starting efficiency of 80%.

The motor flow, $Q_m = 0.785 \times 200 = 157 L / min$

13.5 Volume required/cycle

Motor volume displaced/cycle, $V_m = \dfrac{D_m \times N \times t_i}{60} = \dfrac{0.785 \times 200 \times 0.1}{60} = 0.26L$

Actuator volume displaced/cycle, $V_A = 2 \times (A \times l) = 2 \times 10^{-2} \times 50 \times 10^{-3} = 1L$ (assuming the same actuator area for retract and extend)

Total volume displaced/cycle = 1.26L and for a cycle time of 1.4s, the average flow is:

$$Q_A = \dfrac{1.26}{1.4} \times 60 = 54 L / min$$

13.6 Accumulator

From equation 7, chapter 6, the accumulator capacity, V_0, required to provide the fluid volume ΔV,

$$V_0 = \dfrac{P_2}{P_0} \dfrac{\Delta V}{\left[\left(\dfrac{P_2}{P_1}\right)^{1/\gamma} - 1\right]}$$

Performance Analysis

Taking $P_1 = 100$bar, $P_0 = 0.9P_1 = 90$bar and $P_2 = 150$bar. For 1.26L total volume required in 0.3s we get for the accumulator volume required:

$$\Delta V = 1.26 - \frac{0.3Q_A}{60} = 0.99L \text{ which gives } V_0 = 6.3L \text{ for } \gamma = 1.75$$

The value for γ has been taken from Figure 3, chapter 6 for the application pressure level and the lowest likely temperature.

13.7 Circuit

A basic circuit for this application is shown in Figure 11. Proportional valves would be used for the three flow control functions, their opening being set to give the desired flow in meter-in or meter-out as appropriate. A single valve could be used to operate the two actuators, if preferred, as they are mechanically linked together by the platen. The selected accumulator needs to be capable of providing the maximum flow of 300L/min for closing the platens. The use of an accumulator has provided a low cost solution to this problem as it employs a fixed displacement pump the size of which would have to be slightly greater than the calculated value in order to account for the volumetric efficiency of the pump and motor.

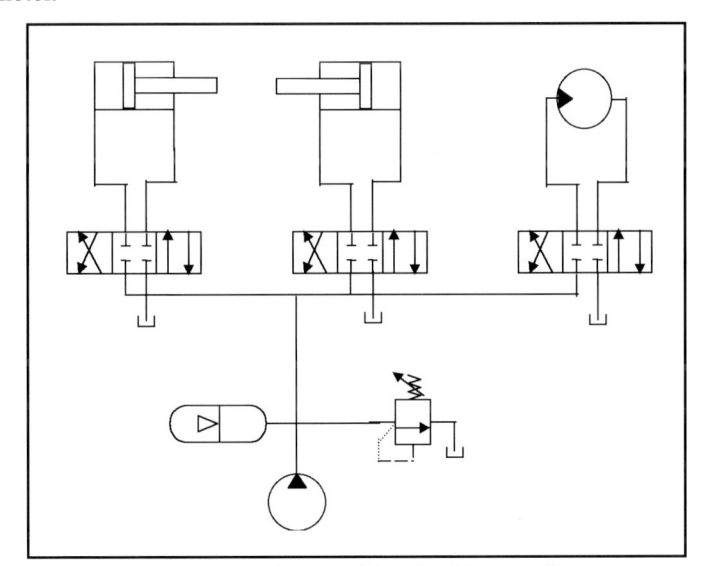

Figure 11. Basic Injection Moulding Machine Circuit

13.8 The prevention of pressure shocks

The use of controlled opening of the valve to reduce pressure shocks was discussed in chapter 10, 4.7 and in this example the response of pressure and actuator displacement to step and ramp valve movements have been evaluated using a computer simulation of the system. The results are shown in Figures 12 for a step change and a 19 ms ramp change in valve movement.

It can be seen that the peak pressure amplitude has been reduced from 60 to 10 bar which represents an 83% reduction in the acceleration force imparted to the actuator and, consequently to the machine itself. Neither of these pressures is in any way excessive but, as can be seen from the actuator displacement, high acceleration (hence, high pressure) is not required to achieve the velocity for sufficient time to move the actuator the specified distance. Indeed, only a very low pressure is needed to maintain the actuator velocity constant against a friction force that has been assumed to be proportional to velocity.

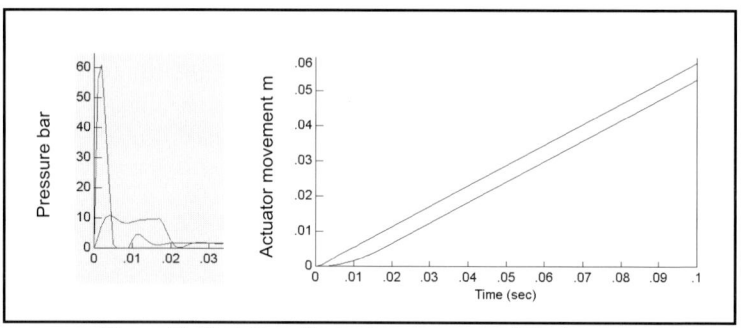

Figure 12. Variation of actuator pressure and displacement

The natural frequency of the hydraulic system can be obtained from the following data:

$A = 10^{-2} m^2$, $m = 250 kg$, viscous damping coefficient = $3000 N/(m/s)$, $\beta = 1.4 \times 10^9 N/m^2$, Volume = $10^{-3} m^3$

$$\therefore \omega_n = \sqrt{\frac{\beta A^2}{Vm}} = 750 \text{rad}/s = 119 Hz \quad \text{Thus } t_c = \frac{1}{119} = 0.0084s$$

The ratio between the ramp time of 19ms to the time for one cycle, t_c is 2.3 which gives a guide to setting the valve opening time to reduce pressure shocks.

Performance Analysis 211

From Figure 12 it is seen that for the valve ramp case the actuator takes about 10ms longer to reach the specified extension of 50mm. The use of a proportional valve to control the flow would allow the ramp time to be adjusted on installation of the system to obtain low-pressure shocks during operation.

14. Oil Cooling

The hydraulic circuit shown in Figure 13 is used to provide the feed of a circular saw and employs a meter-out pressure compensated flow control valve with a fixed displacement pump. It is required to select an appropriate cooler from the performance data in Figure 14. Assume that there is no heat loss to the surroundings.

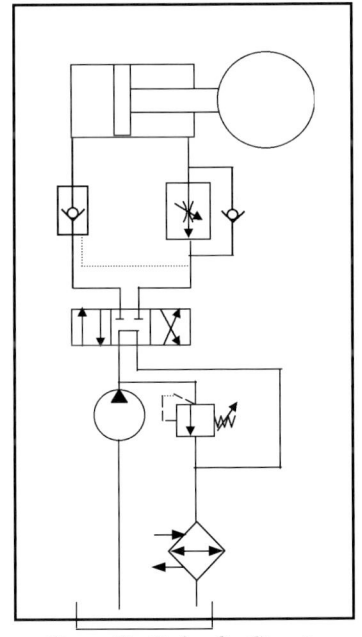

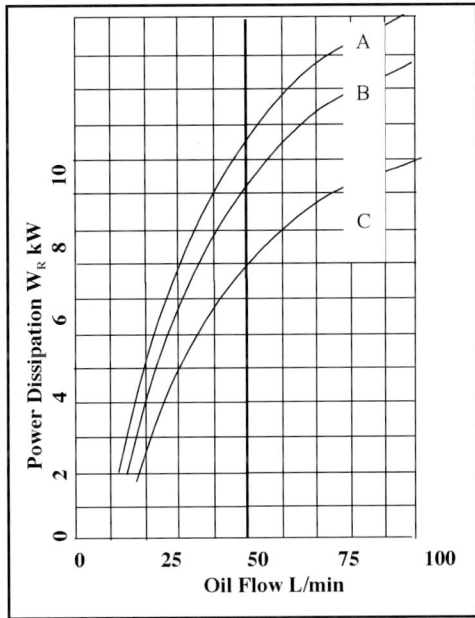

Figure 13. Hydraulic Circuit   *Figure 14. Cooler Performance Characteristics*

14.1 System data

Pump flow Q_S = 50L/min
Relief valve pressure P_S = 200bar
Actuator area ratio a_R = 1.33

Pressure losses between pump
and actuator (both sides) ΔP_L = 20bar = Pressure loss for valve in central position
Open centre valve pressure loss = 20bar
Cooler performance conditions:
Oil to water inlet temperature difference $(\Delta T_{OW}) = 40^0C$ for rated cooling power (w_R)

Performance at other temperature differences $W = W_R \dfrac{\Delta T_{OW}}{40}$ kW

Water flow = 1.4L/min/kW

14.2 Duty cycle

Actuator extend

Duration t_E = 120s
Piston flow Q_E = 20L/min
Relief valve flow Q_V = 30L/min
Actuator return flow Q_R = 15L/min

Actuator retract

Duration t_R = 36s
Annulus flow Q_A = 50L/min
Piston flow Q_P = 66.7L/min

Idle condition

Duration t_I = 60s
Flow Q_S = 50L/min

14.3 Heat generated

a) Actuator extension

The actuator extends under the control of the meter-out pressure compensated flow control valve so that the pressure created at the actuator outlet (annulus end) will rise to provide a force that satisfies the force balance on the actuator piston. The reaction force from the saw will normally be in opposition so the highest annulus pressure will arise when this force is zero.

Performance Analysis

For this situation, the annulus pressure $P_A = P_P \times a_R = 200 \times 1.33 = 240\,bar$. This pressure will be dissipated as heat in the flow control valve and the pipes as will the pipe pressure loss in the inlet flow from the pump.

Energy dissipated in heat $\dfrac{Q(L/min)}{60 \times 1000} \times P(bar) \times 10^5 \times t(s) = 1.67QPt$

Energy dissipated in the inlet flow $H_E = 1.67 Q_E \Delta P_L t_E = 1.67 \times 20 \times 20 \times 120 = 80\,kJ$
Energy dissipated in the return flow $H_R = 1.67 Q_R P_A t_E = 720\,kJ$
Energy dissipated in the relief valve flow $H_V = 1.67 Q_V P_S t_E = 1200\,kJ$

b) Heat generated during retraction of the actuator

Energy dissipated in the inlet flow $H_E = 1.67 Q_A \Delta P_L t_R = 60\,kJ$
Energy dissipated in the return flow $H_R = 1.67 Q_P \Delta P_L t_R = 80\,kJ$

c) Heat generated during idle period

Energy dissipated through open centre valve $H_I = 1.67 Q_S \Delta P_L t_I = 100\,kJ$

d) Total heat generation and cooler selection

The total heat generation = 2240kJ

Average power dissipation = $\dfrac{2240}{216} = 10.4\,kW$

For this power to be dissipated it is necessary to select an appropriate cooler from Figure 14. The heat dissipated by the coolers is based on a difference between the oil and water inlet temperatures of 40°C and a water flow of 1.4L/min/kW. The mean oil inlet temperature will need, therefore, to rise so as to create a sufficient temperature difference for the required power dissipation.

The power dissipation (W) at any other temperature difference is given by:

$$W = W_R \dfrac{(T_I - T_W)}{40}$$ where W_R is the power dissipation at

an oil flow of 50L/min, T_I the oil inlet temperature and T_W the water inlet temperature. For coolers A and B the power dissipation at different oil inlet temperatures is shown in Table 5 for a water inlet temperature of 20°C.

Oil inlet temperature $T_1\ ^0C$	Power dissipation W kW	
	Cooler A (W_R = 11.5 kW)	Cooler B (W_R = 10.1 kW)
50	8.6	7.6
55	10	8.8
60	11.5	10.1

Table 1. Power dissipated in the cooler

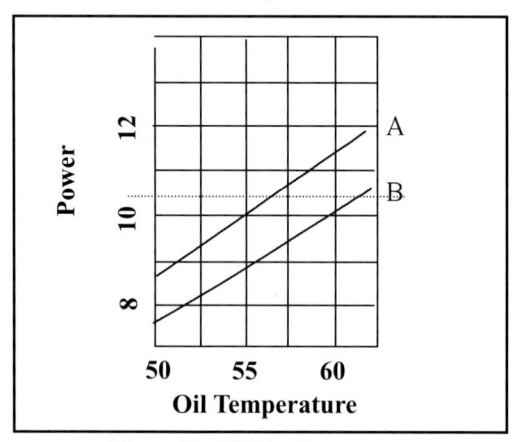

Figure 15. Oil Inlet Temperature

The variation in the cooler power of Table 1 can be used to obtain the operating oil inlet temperature as shown in Figure 15. To dissipate 10.4kW cooler A will have to operate at an inlet temperature of 58⁰C and for cooler B, 63⁰C will be required. The reservoir temperature can be determined from:

$$W = \rho C_P Q \Delta T_C$$

$$\therefore \Delta T_C = \frac{W}{\rho C_P Q} = \frac{10400 \times 60 \times 1000}{870 \times 2100 \times 50} = 6.8^0 C$$

Thus the reservoir temperatures will be respectively, 51.2 and 56.2⁰C.

The required water flow = *1.4 x 10.4 = 14.6L / min*

14.4 Heat loss to the surroundings

The heat loss to the surroundings is difficult to quantify because of the wide range of heat transfer coefficients that can apply. If we consider a cube shaped reservoir having a capacity of 150L with the dimensions of 0.6m the surface area will be 2.16m².

The power dissipation due to convection is given by $W = UA\Delta T_A$ where U is the heat transfer coefficient that can have values in the range 2 - 25 W/m²°C. An effective area, A, of 2m² with a mean tank temperature, T_T, of, say, 50°C, and an ambient temperature, T_A, of 20°C, would provide heat dissipation in the range:

$$W = 2 \times 2 \times 30 = 120W \text{ to } 25 \times 2 \times 30 = 1.5kW$$

Consequently, heat dissipation to the environment can be quite significant but unreliable because of its variation with ambient conditions. It is normal to incorporate a thermostat into the cooling system that operates a flow control valve in the water supply so that the effects of ambient temperature and load force variations on the oil temperature can be reduced.

14.5 Pump efficiency

The efficiency of the pump will also create a heat load in the oil. Thus for a pump overall efficiency of 90%, the heat generated in the pump due to this inefficiency is:

$$(1 - \eta) \times \frac{Q(L/min)P(bar) \times 10^5}{60 \times 1000 \times 1000} = 0.1 \frac{QP}{600} kW = 1.67 kW$$

Using a variable displacement pressure compensated pump will reduce the heat generated in the pump but more importantly, reduce the heat generated in the relief valve because this flow will be eliminated thus approximately halving the heat dissipation. The size of the cooler could therefore be reduced by approximately 50% as can the required water flow.

15. Vehicle Transmission

A track driven earth-moving vehicle uses a hydrostatic motor in each caterpillar track drive, the hydraulic power being supplied from a variable capacity pump driven at constant speed by the engine.

For the vehicle climbing an incline of 20 degrees, after making suitable allowance for losses, estimate:
a) the maximum vehicle speed.
b) the fluid pressure and flow in the transmission.
c) the effect on the transmission efficiency when the fluid viscosity reduces to 20 cSt.
Data:

Maximum engine power	250kW
Hydraulic motor capacity	2000 cm³/rev
Motor gearbox ratio	2:1 reduction
Relief valve setting	280 bar
Vehicle wheel diameter	0.55 m
Vehicle weight	270 kN
Friction (speed independent)	10 kN
(speed dependent)	4 kN/(m/s)
Track drive efficiency	90%
Motor/gearbox mechanical efficiency	92%
Pump/Motor volumetric efficiency	95% (at 35 cSt viscosity)
Pump mechanical efficiency	95%

a) Vehicle speed

Total load force at the vehicle tracks = *270sin 20° + 10 + (4U)kN*

The maximum vehicle speed may be limited by the available power, the system pressure or the system flow rate. In this case, the maximum speed is limited by the power available.

The overall efficiency = *0.90 x 0.92 x 0.95 x 0.95 x 0.95 = 0.71.*
The output power at the tracks = *0.71 x 250 = 178kW*
Hence the maximum vehicle speed whilst climbing the *20°* gradient is given by:

$$178 = (270 \sin 20° + 10 + 4U)U \text{ kW}$$

$$\therefore 4U^2 + 102.3U - 178 = 0$$

For which $$U = \frac{-102.3 \pm \sqrt{102.3^2 + 2.9 \times 10^3}}{8} = 1.64 \text{ ms}^{-1}$$

b) Pressure and flow

For the maximum vehicle velocity, the maximum load force is given by:
Max force at the tracks = 270 sin *20°* + 10 + *(4 x 1.64)* = 109kN

Performance Analysis

The motor torque is given by: $Torque = \dfrac{Max\ Force}{2 \times 2} \times Wheel\ radius$ (i.e 2 motors and a gearbox reduction ratio of 2:1).

$$\therefore T = \dfrac{109}{4} \times \dfrac{0.55}{2} = 7.5\ kNm$$

For this torque the ideal system pressure is:

$$P_{ideal} = \dfrac{T}{D} = \dfrac{7.5 \times 10^3}{\left(\dfrac{2000 \times 10^{-6}}{2\pi}\right)} = 236\ bar$$

And for the motor/gearbox mechanical efficiency of 92%, and the track drive efficiency of 90% the maximum working pressure is:

$$P = \dfrac{236}{0.90 \times 0.92} = 285\ bar$$

For the vehicle speed of *1.64ms⁻¹*, the ideal flow rate for all the motors is given by:

$$Q_{ideal} = 4(DN) = 4\left[2000 \times 10^{-3}\left(\dfrac{1.64 \times 60}{\pi\, 0.55}\right)\right] = 455\ L min^{-1}$$

Therefore, allowing for the motor volumetric efficiency, the maximum flow required at the motor inlet ports is:

$$Q = \dfrac{455}{0.95} = 479\ L/min$$

Power check:
Hydraulic Power

$$= P \times Q = 285 \times 10^5 \times \dfrac{479 \times 10^{-3}}{60}\ (= \dfrac{PQ}{600}) = 227.5\ kW$$

Pump input power

$$= \dfrac{227.5}{0.95 \times 0.95} = 252\ kW$$

c) The effect of fluid viscosity

Motor $\eta_v = 0.95$ at 35 cSt

$$\text{Now} \quad \eta_V = \frac{Q_t - Q_s}{Q_t} = 1 - \frac{Q_s}{Q_t} \quad \therefore \frac{Q_s}{Q_t} = 0.05 \quad \text{(refer to chapter 9)}$$

$$\text{For a fluid viscosity of 20 cSt} \quad \frac{Q_s}{Q_t} = 0.05 \times \frac{35}{20} = 0.0875$$

$$\therefore \eta_V = 91.25\%$$

Assuming that the pump volumetric efficiency will change by the same proportion with viscosity as the motor.

Overall efficiency at 35 cSt = 71%. At 20 cSt, the overall efficiency

$$= 71 \times \left(\frac{91.25}{95}\right)^2 = 65.5\%$$

d) Pump controls

The methods of controlling the flow and pressure from the variable displacement pump discussed in chapter 7 can be applied to this transmission system. It is required to control:

- Maximum pressure.
- Pump torque. At a constant pump speed this would limit the input power to the pump.
- Pump output flow.

Figure 27 of chapter 7 is reproduced below as Figure 16, which portrays the three parameters that are to be controlled. Maximum pump pressure is controlled by the compensator shown in Figure 17 (reproduced from Figure 25, chapter 7) that reduces the pump displacement should any service demand a pressure that is higher than that set by the compensator (valve A). This control is capable of reducing the net output flow to zero if necessary (stalled condition).

Performance Analysis

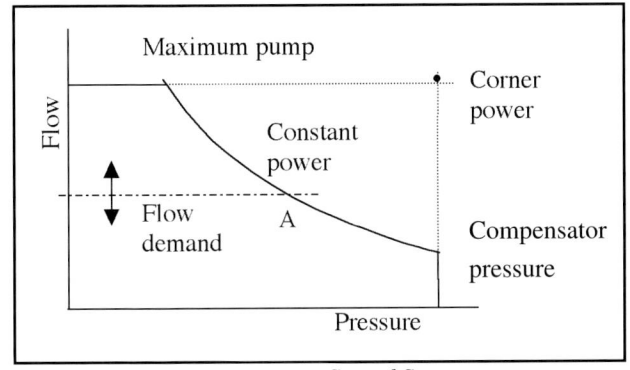

Figure 16. Pump Control Strategy

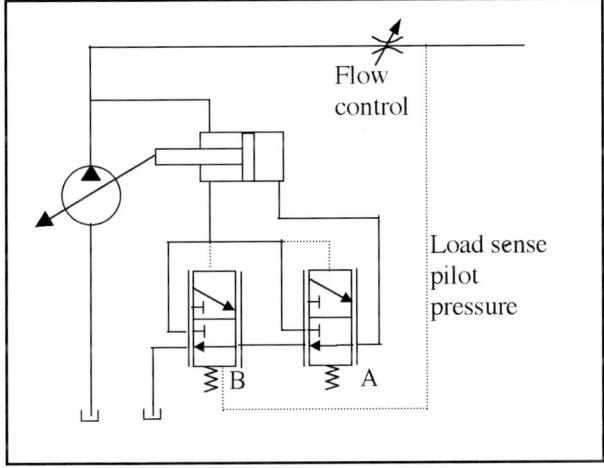

Figure 17. Variable Displacement Pump with Pressure Compensation and Flow Control

The pressure drop across the selected valve opening will control the pump displacement to maintain constant flow to the service (valve A). Referring to Figure 16 this sets the flow at 'A' which will be limited by either:

- the pressure compensator, as described above or
- the setting of the pump power (torque) control shown in Figure 18 (reproduced from Figure 26 chapter 7).

The operation of the power control, which is described in chapter 7, is achieved by sensing a pressure signal that varies with the pump displacement. This pressure provides a feedback to the pump displacement control valve that will maintain the displacement so that the pump output power is constant under changing load pressure conditions. An analytical method for evaluating the steady state performance of the control is given at the end of this section.

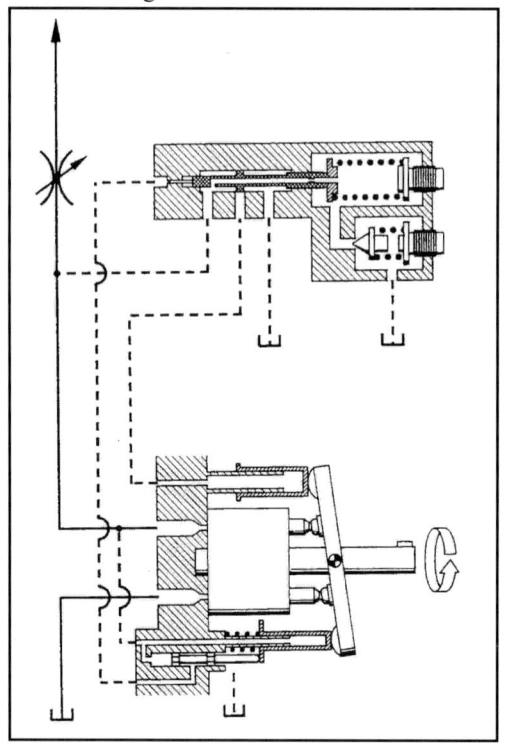

Figure 18. Pump Power (torque) Control

For the track drive with the vehicle travelling on flat ground against low resistance, maximum speed will be obtained because the required pressure will be reduced in relation to that required when moving up the 20^0 gradient.

Referring to Figure 16, the flow of 479L/min, (b), will be selected by the flow control valve the pump displacement being controlled by the pressure drop valve B in Figure 17. The pressure of 285 bar that is required for the 20^0 incline may be limited by the constant power control (point A at, say, 15% lower pressure than the pump compensator setting).

With a displacement increase of, say, 150% (2.5:1), the vehicle will be capable of travelling at 2.5 x 1.64 = 4.1 m/s (14.8km/hr). The higher flow will be selected by the flow control valve the pressure being limited by the constant power control should the pressure increase above the corresponding value for the selected flow. Maximum speed will naturally be determined by the maximum displacement of the pump.

Steady state pump power control analysis (Figure 18)

Displacement control valve

The steady state valve forces are given by:

$$P_s a_1 + P_p a_2 = C + KX$$

For a spring stiffness K, preload C pressure sensing areas a_1 and a_2. X is the valve movement away from the closed position.
For $X = O$ in steady state conditions the force balance of the valve gives:

$$P_S = \frac{C}{a_1} - P_P \frac{a_2}{a_1} \quad (1)$$

This gives a relationship between the two pressures for the valve to be held in a closed position ($X = O$).

Displacement sensing hydraulic potentiometer

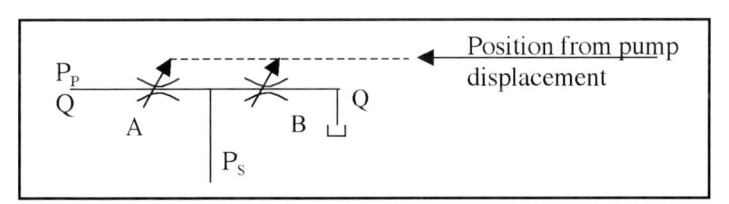

The length of the restriction A (L_1) is reduced with increasing displacement and the length of B (L_2) increases. Considering the flow through the pressure signal hydraulic potentiometer and assuming laminar flow (i.e. $Q \propto \frac{\Delta P}{L}$) gives:

$$Q = \frac{K_1(P_P - P_S)}{L_1} = \frac{K_1 P_S}{L_2} \quad (2)$$

Now $L_1 + L_2 = L$ (Total length of flow path). (3)
  $L_1 = O$ when pump displacement $D = D_{max}$
  $L_2 = O$ when pump displacement $D = 0$

Choosing the geometry such that:

$$L_1 = L(1 - \frac{D}{D_{max}}) \quad \text{and} \quad L_2 = L\frac{D}{D_{max}} \quad (4)$$

Equation (2) gives:

$$P_S = \frac{P_P}{1 + L_1/L_2} \quad (5)$$

Then, with equation (4)

$$P_S = \frac{D}{D_{max}} P_P \quad (6)$$

Equations (1) and (6) give:

$$P_S = \frac{D}{D_{max}} P_P = \frac{C}{a_1} - \frac{a_2}{a_1} P_P \text{ , so } P_P(\frac{D}{D_{max}} + \frac{a_2}{a_1}) = \frac{C}{a_1} \quad (7)$$

Equation (7) shows that in the steady state, increases in the pump pressure P_p will be associated with reductions in the pump displacement D at a level that depends on the valve sensing area ratio to give an approximate constant power with the pump operating at constant speed.

16. Pump Control Applications

16.1 Application of Pump Unloader Valve to Vehicle Crusher/Refuse Machines

A pump-unloading valve can be used as shown in Figure 19 so that high speed can be obtained from both pumps together. When the force increases so that the pressure rises to a level that is higher than the setting of the unloading valve the valve opens and allows flow from one pump to be bypassed at low pressure to tank. The check valve separates the two circuits.

To limit the maximum power the set pressures for the unloading and the relief valve need to be in the ratio:

$$\frac{P_2}{P_1} = \frac{Q_T}{Q_2} = \frac{Q_1 + Q_2}{Q_2} = 1 + \frac{Q_1}{Q_2}$$

Performance Analysis   223

Thus for $Q_2 = \dfrac{Q_1}{2}$; $\dfrac{P_2}{P_1} = 3$.

Triple pumps can provide three pressure/flow ranges by the use of two unloading valves set at different pressures.

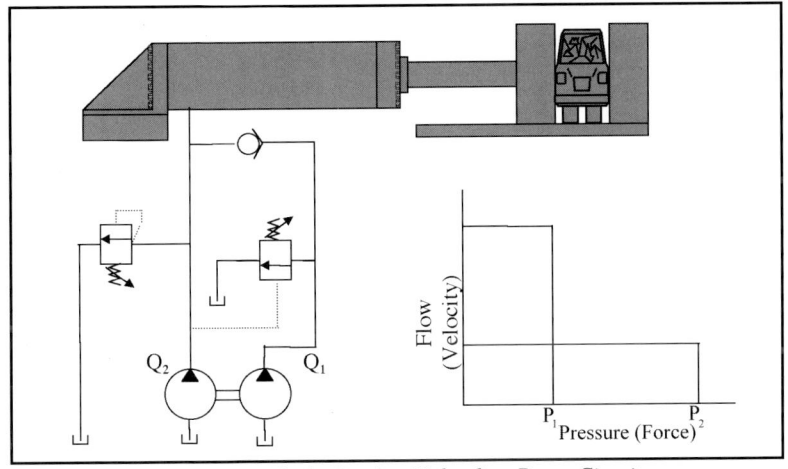

Figure 19. Vehicle Crusher Unloading Pump Circuit

16.2 Application of Pump Controls to Bending Machines

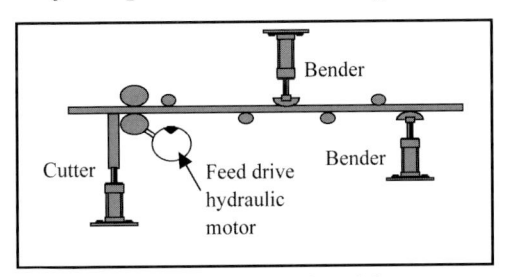

Figure 20. Bending Machine Schematic

Operation of the bending machine requires the following actions to be carried out:

- Rotate drive rollers (hydraulic motor)
- Operate cutter
- Operate two bender actuators

Push finished piece into a skip

Figure 21. Bending Machine Circuit

The circuit in Figure 21 controls the flow to each service by maintaining the pump pressure at a level set by valve B from the highest load pressure. The directional control valves set the flow required for each service with the individual flow compensators maintaining constant pressure drop across each valve.

Note: A shuttle valve is often used to select the highest actuator or motor pressure for the load sense signal as shown in Figure 22:

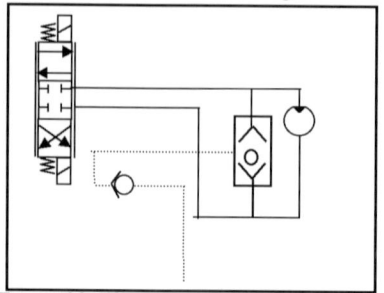

Figure 22. Alternative load sensing system

Performance Analysis

17 Application of compensation techniques

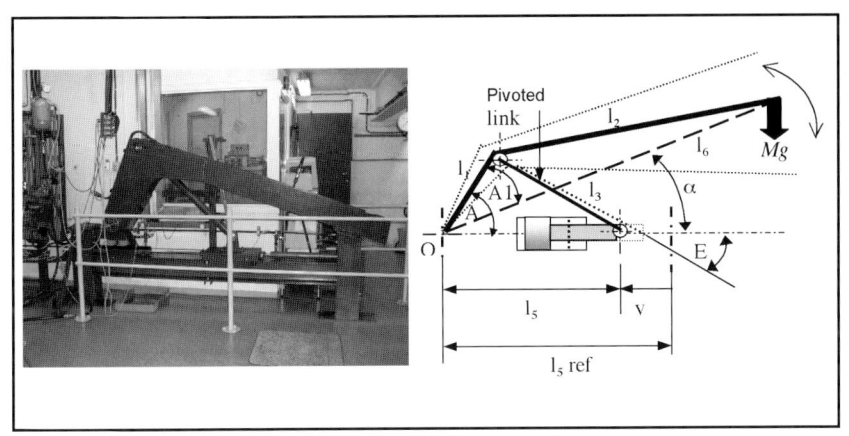

Figure 23. Hydraulically operated crane

It is required to improve the performance of the electrohydraulically operated crane mechanism shown in Figure 23, which is controlled by closed loop position feedback. The compensation techniques discussed in Chapter 10 will be used to demonstrate the application of these methods.

17.1 Steady state accuracy

A major problem experienced by the crane is the effect of changes in the external forces on the steady state position due to valve leakage when it is in the null

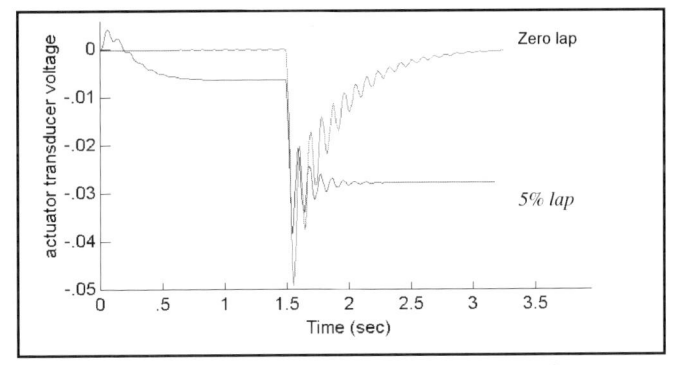

Figure 24. The effect of changes in external forces on steady state errors

position. A result obtained from a simulation of the system is shown in Figure 24, which has a feedback transducer gain of *20 V/m*. This shows that for the change in the force there will be an error of about *1.5mm* in the crane position for the valve having leakage at the null position (underlap).

17.2 °

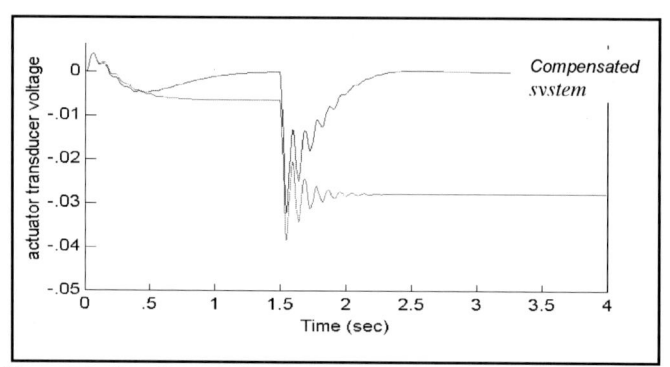

Figure 25. The effect of adding proportional plus integral compensation

Figure 25 shows how the use of integral plus proportional (*P+I*) compensation removes the steady state error that arose in the original system with the application of external forces. The choice of coefficients for the compensator depends on their effectiveness and the stability of the overall system.

The 2nd method of Ziegler-Nichols for selecting the gains is based on the value of the loop gain, K_M, that will cause continuous oscillations of the system and the frequency of the oscillations, f_n. Continuous oscillations occur at a natural frequency of *10Hz* with an open loop gain value, K_M of 6. which creates for the crane simulation we get:

The *P+ I* compensator has the transfer function given by:

$$(P + I) \text{ Transfer functions} = K_p(1 + \frac{s}{T_I})$$

Where the coefficient values by the Ziegler-Nichols method are given by:

$K_p = 0.45 K_M$ and $T_I = 0.83 T_p$ and T_p = periodic time of oscillations = $1/f_n$

Figure 26 shows the system step response with different values of the integral coefficient *a* where it can be seen that the value has an important effect on the overshoot and successive oscillations.

Performance Analysis

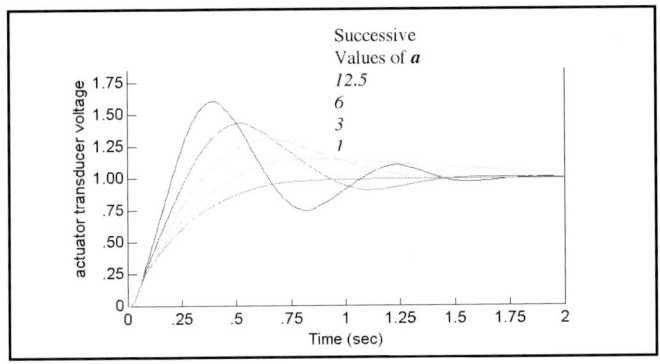

Figure 26. Step response with the P+I compensator

The values obtained using the Ziegler-Nichols method give $K_p = 2.7$ and $T_I = 0.083$ and a transfer function:

$$\frac{2.7}{s}(s + 12.5) \qquad (1)$$

Or $\quad = \frac{2.7}{s}(s + a)$ (for the uncompensated system $a = 0$)

These values are those used in the simulation, the results of which are shown in Figure 26.

17.3 Proportional plus integral plus derivative (PID) compensation

The transfer function for the *PID* compensator can be written:

$$G_C = K_p \left(1 + \frac{1}{T_I s} + T_D s\right) \qquad (2)$$

The Ziegler-Nichols 2nd method proposes the following values of the *PID* coefficients:
- $K_p = 0.6 \times$ the proportional gain that just causes the system to oscillate
- $T_p = 0.5 \times$ the time for one cycle of the oscillations $= \dfrac{0.5}{f_n}$

Applying the values from the table for the crane linearised simulation with the data from section 2 gives a transfer function for the *PID* compensator:

$$G_C = 3.6 \left(1 + \frac{1}{0.05s} + 0.013s\right) = \frac{3.6}{s}(0.013s^2 + 20) \qquad (3)$$

For the system the natural hydraulic frequency is *10 Hz* with a damping factor of approximately *0.6*. Using these values gives a linearised transfer function of the system:

$$Kp \frac{1}{(1 + 0.002s + 0.00025s^2)} \qquad (4)$$

The value of K_p is set in the controller amplifier, which for applying the *PID* compensator will be *3.6*.

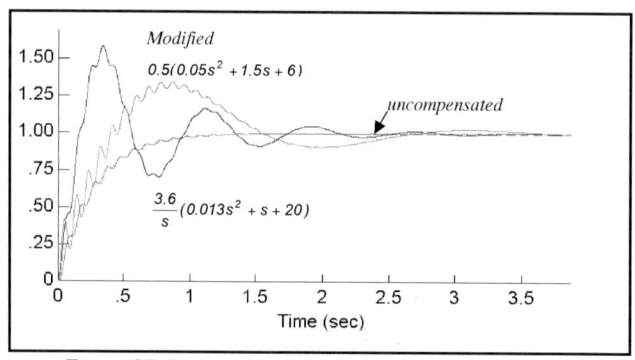

Figure 27. Dynamic response using a PID compensator

17.4 Load pressure feedback

The application of load pressure feedback was shown to improve the damping coefficient of the closed loop hydraulic system. The effect of increasing the damping ratio by ten times to around *0.6* allows the system gain to be increased by more than *100%* as shown in Figure 28.

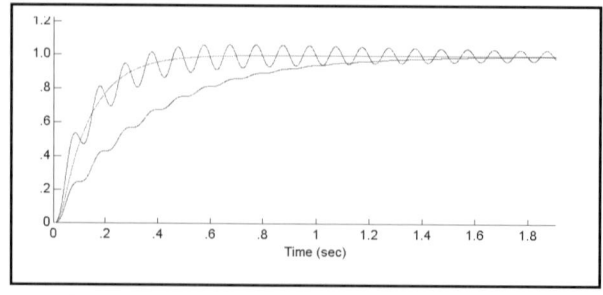

Figure 28 The application of load pressure feedback

The increased damping ratio is seen to not only allow the use of an increase in the system gain but also eliminates oscillations even at the higher gain value.

17.5 Concluding remarks

The results using the Ziegler-Nichols values for the coefficients give excessive oscillations for both the *P+I* and *PID* compensators and changes to these values provided considerable improvements. Both of these methods provide the elimination of steady state errors and improved response but optimisation to reduce oscillations requires a trial and error approach which is greatly assisted by the use of simulation techniques.

This emphasises the considerable choice that is available for setting these gains in that stability is obtained but the damping is not clearly defined. In most systems the values of the coefficients are determined by trial and error methods.

Other methods are also available mostly using digital techniques, which include observer, pole placement, sliding mode control and other forms of robust control.

CHAPTER TWELVE

SYSTEMS MANAGEMENT

12. SYSTEMS MANAGEMENT

Summary

The management of a hydraulic system is concerned with all aspects from the initial design stage to its final disposal after completing a useful life. During the design phase the components and circuits that are selected for the system need to provide the performance that is required to meet the machine specification and have the necessary life to meet the expected duty cycle under the given environmental conditions. It is also important to ensure that the system is properly installed and commissioned and that appropriate maintenance and operational procedures are put into place by the machine builder/user.

Ultimately the major aspect of management is to achieve the most economic cost over the total lifetime of the equipment. This does not refer only to the cost of purchase because a higher cost system may provide a lower running cost so that in its lifetime the total cost may be less than that obtained with a system having a lower purchase cost. Reliability is a major issue because the level of this will reflect on the maintenance of the system and its operating cost in relation to the cost of machine 'downtime'.

It may be preferable in some circumstances to use condition-monitoring techniques to determine loss of performance and predict component failure (see *Coxmoor Publishing's Machine & Systems Condition Monitoring Series* of handbooks). This may be required where a high level of safety is demanded by the specification and often it may be necessary to perform some form of fault analysis for the system. All of these aspects will need to be considered as part of the systems management process.

1. Introduction

The management of hydraulic systems is an important feature that needs to be considered during the design process, the level of which will depend on the ultimate use of the system. Clearly this involves a number of factors which include the complexity of the system, the machine duty cycle and its environment and the level of reliability required by the machine builder/user.

2. Aspects of systems management

In general systems management will need to:

- Ensure that the proposed system will meet the machine specification.
- Carry out a fault or failure analysis of the system.
- Establish that the procured components/systems achieve the specified levels of cleanliness and are protected prior to, and during, installation from the ingression of contaminants.
- Check that the system has been installed correctly, flushed and filled to the correct level and commissioned appropriately.
- Determine the type of maintenance procedure (i.e. preventative or corrective) to be used.
- Establish the training of personnel to a technical level of competence in order to achieve reliable and economic operation of the system and deal with problems that arise during its use

3. Systems management objectives

The necessity and level of systems management needs to be determined. For example systems used for research test rigs, complex manufacturing machines and auxiliary drives on mobile plant will require quite different approaches. The objectives of the management system will, therefore, need to be defined. Likely objectives would include those concerned with:

- The minimisation of operating costs.
- The maintenance of end product quality.
- The maximisation of system reliability.
- The assurance of a high level of safety.

The achievement of these objectives will be strongly related to the design of the system in relation to the machine duty cycle and its environment. To achieve high reliability it may be considered appropriate to measure major parameters that would include pressures, temperatures, valve positions, pump and motor

Systems Management   235

displacements and flows for condition-monitoring purposes. This would allow predictive maintenance procedures to be used, which can avoid machine downtime due to component failures. It has been shown that reductions in the total cost of systems in their lifetime can be upwards of 10% by the use of condition-monitoring methods.

4. System cleanliness

The premature failure of components and unreliable operation of hydraulic systems is often the result of inadequate contamination control, the incorporation of which should be part of the system design process. During installation of the system the levels of particulate contamination should be monitored using a particle counter and the system should be operated until these levels have reduced to the specified value. This process may require the filter to be changed in order that contaminants do not pass into sensitive components. When the contaminant levels have reached a satisfactory level the system should be drained and a new filter fitted.

Ideally the achievement of optimum system life and reliability is obtained by co-operation of the various manufacturers/suppliers involved in its manufacture and operation which is represented in Figure 1. The application of such a total cleanliness control programme will reduce the failure rate for systems as shown by the well-known 'Bath-tub' curve in Figure 2. This has particular effect on the number of failures at the beginning and end of the product life and consequently the overall life.

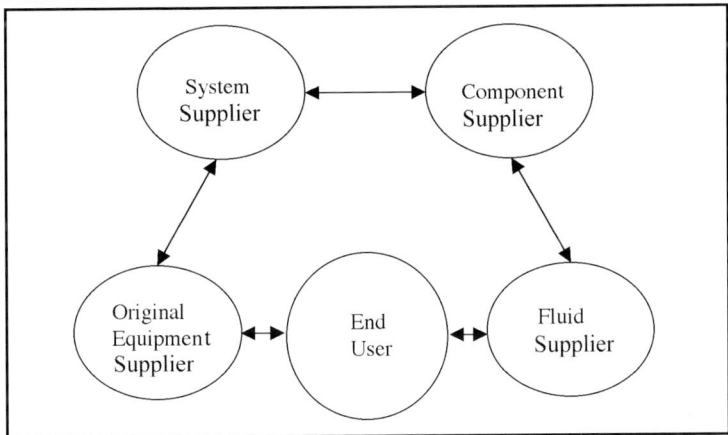

Figure 1. The Complete Partnership

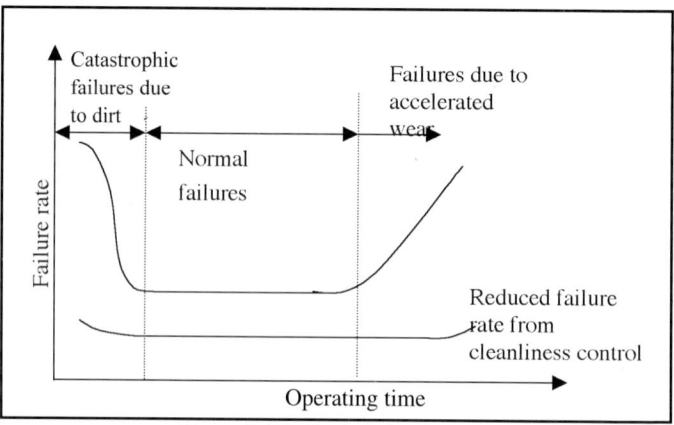

Figure 2. The Bath-Tub Life Curve

5. Fault analysis

The evaluation of circuit faults needs to examine all of the system operating modes and states. However, circuit drawings are often complex, they are only able to show one operating state and, generally, they do not show component sizes, operating conditions and ratings.

Two commonly employed fault finding methods include Fault Tree Analysis and Failure Modes Effects Analysis (FMEA).

5.1 Fault Tree Analysis (FTA)

Fault trees are drawn to show possible causes of major malfunctions of the system that usually deals only with those failures that are considered to be most likely to occur, events of low probability being excluded.

This approach can be applied to the actuator circuit shown in Figure 3 for considering all the possible failures that will prevent movement of the actuator. Most of the causes of this failure are single events, which would include:

- DCV not opening OR
- No pump flow (e.g. drive motor failed, inlet suction filter blocked) OR
- Relief valve failed open (e.g. broken spring) OR
- Load force greater than that available from the actuator OR
- Insufficient fluid in the reservoir

Systems Management

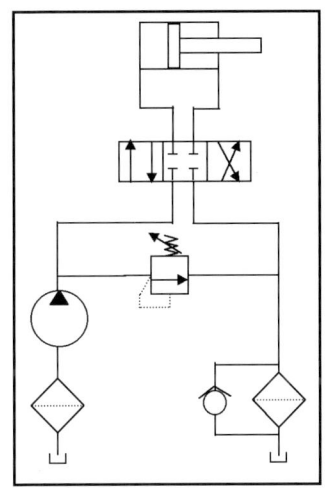

Figure 3. Valve Actuator System

Some faults require double failures to occur such as blockage of the return line filter <u>AND</u> the bypass valve jammed closed. This is normally an extremely unlikely situation but in some systems where safety interlocks are used these types of failures might need to be considered. A typical example is the circuits used for the control of hydraulic presses.

Thus a fault tree can be drawn by following the circuit from the component associated with the top event along the lines that lead either to the supply (pressure) or the return (tank), the top event itself may be linked to more basic faults. The event statements are linked through logic gates. OR gates require only one input to be available before the output event occurs. AND gates on the other hand require both input events to have occurred before the output event can happen.

In the simplest circuits, only OR gates are required, linking alternative fault events, e.g. there may be no flow from a directional control valve due to no input flow, OR failure of the pilot signal selecting the valve open position, OR valve jammed shut. In circuits with in-built redundancy, AND gates are also required, e.g. there is no flow in a specified line when there is no flow from the main pump supply AND no flow from the auxiliary pump supply.

5.2 Failure modes effects analysis (FMEA)

The fault tree analysis works from the top down in that a key malfunction is selected first and failures that can cause this are then established. The FMEA method works from the bottom up in that it determines the effect on the system of every failure mode of every component. This is a long and tedious procedure that is very suited to computerised methods.

An advantage that this method has over the FTA is that all failures are identified, some of which may have been missed in the FTA. However, identifying the relative importance of the failures is not part of the process and to reduce the amount of work involved the Pareto method is often used wherein only major malfunctions are considered, as in the FTA. When carrying out these analyses the fluid, the effects of the environment and the possibility of operator error must also be included.

If failure probability data is available, usually in terms of Mean Time Between Failures (MTBF), the reliability of the system can be established. However, such data for commercially available components does not usually exist and mostly this approach can only be applied to military projects and military/civil aircraft where it is necessary to perform a quantitative analysis of the system reliability.

6. General

Systems management involves some or all of the activities that have been discussed, which approach that is used depending on the requirements of the machine to which it is being applied. It may be considered on the basis of cost and/or safety to use condition-monitoring techniques for managing the operation of the machine. The parameters that can be monitored in fluid power applications include:

- Pressure
- Flow
- Temperature
- Torque
- Energy consumption
- Contamination
- Vibration
- Acoustic emission
- Noise
- Speed
- Position (valve, actuator)

The most appropriate monitoring is determined by undertaking a careful examination of the failure modes and the available measurement methods. The priority methods would be those required to provide information on the most likely failures and then, depending on the importance of the system, other methods can be incorporated if necessary.

INDEX

Index

A

accumulator(s), 63-6, 104, 208
  charging circuit, 101
  volume, 209
actuator(s)
  circuit, 82
  control systems, 137
  cushion, 194
  double-acting, 30
  fatigue life of, 34
  force control, 162
  hydrostatic transmission circuit, linear, 104
  linear, 44
  mountings, 31, 170
  performance, rotary, 38
  selection, 33
  single ended, 177
  systems, 105, 171
  tie rod construction, of 31
  transmissions, linear, 103
  types, 37
  valve, 154, 157
  vane rotary, 38
adiabatic index, 67
adjustable restrictor valve, 176
aeration, 108, 114
air release, 74
ancillary equipment, 63
annular ports, 88
annulus, outlet metering, 122
asymmetrical metering, 95
axial
  momentum, 120
  piston,
    bent axis type of, 15
    motor, 15

B

ball check valve, 82
Bath-Tub life curve, 236
bending machine circuit, 224
Bernouilli
  force, 120
  Energy Equation, 117
Beta ratio for filters, 70
bleed off control, 95, 86, 97
Bode
  diagram, 152
  plot, 153, 156, 169, 205
boost
  check valve, 171
  flow, 102
  pressure, 23, 103
brake(s) 184
  circuit, motor, 103
  valves, 190
breather, 74
buckling load, 32
bulk modulus, 143-4
bypass
  circuits, 56
  control, 95
  filter, 107
  regulating valve, 97
  valve,
    central, 58, 97, 196
    characteristics, 96
    pressure compensated, 98

C

cam
  motors, radial piston, 22
  type motor, radial piston, 21
cartridge
  relief valve, two stage, 44
  valve, 46
cavitation, problems of, 84
centiStoke (cSt), 112
central bypass valve, 58, 97, 196

circuit(s),
  linear actuator
    hydrostatic transmission, 104
    meter-in, 83
    motor brake, 103
    multiple actuator, 86
    open loop, 16
    options, 77
    pump unloading, 105
    rotary hydrostatic transmission, 103
cleanliness control, 235
closed loop
  control, 21, 58, 98
  systems, 137, 139
  hydraulic system, 144, 228
  performance, 170
  position, 138, 154
  system performance, 161
  systems, 48, 105, 169
coil hysteresis, 150
compensation techniques, 164, 225
component failures, 235
compressed volume, 137
condition-monitoring methods, 235
constant
  power control, 100
  pressure, 42
contaminant levels, 71, 235
Contamination, 238
  control, 67, 78, 106, 235
control,
  constant power, 100
  electrohydraulic, 137
  feed back, 140
  load sensing, 99
  meter-in, 83-4, 86-7
  meter-out, 85, 94, 97, 105
  velocity, of, 102
  open loop velocity, 138

restrictor, 79
secondary, 104
sequence, 106
servo valve, 172
valves,
 load, 46, 105
 pressure, 41, 111
 variable displacement pump, 98
cooler(s), 63, 211, 214
 air, 72
 characteristics, 73
 performance characteristics, 74
 types, 72
 water, 72
counterbalance valve, 47
 systems, 48
cross line relief valves, 189-90
cushion methods, 36
cushioning valves, 38

D

damping
 factor, 169, 205, 228
 ratio, 137, 228
DCV, 82-3, 236
derivative control, 165, 166
diaplacement pump, variable, 53
directional control, 53
 valve, 237
displacement
 controls, pump, 16
 mechanism, pump, 99
 motor, 21, 181-2, 186, 208
 pump, 190, 218-20, 222
  variable, 171
  fixed, 53, 175, 209, 211
double pump system, 106
drain flows, 190
dual relief valves, 44
dynamic
 response, valve
 viscosity, 112

E

eccentric motor, radial piston, 21
edges, notched metering, 97
efficiency, 19, 78, 127
 mechanical, 131-2, 216
 overall, 132
 pump volumetric, 218
 volumetric, 127, 131-2, 209
electro-hydraulic valves, 16
electrohydraulic
 closed loop system, 169
 control, 137
 system, 98
 valve, 171
 servo valve, 55, 171
 systems, 161
 valve, 141, 150
equal area actuator, 95
Euler buckling load, 32
external gear pump, 12, 13

F

failure
 analysis, 234
 component, 233, 235
 modes, 238-9
Failure Modes Effects Analysis (FMEA), 236
fault
 analysis, 233-4, 236
 events, 237
 tree, 237
  analysis, 236-7
feed back control, 140
filter(s), 64, 67-8, 70, 106
 Beta ratio for, 70
 bypass, 107
 circuit(s), 107
  low pressure, 108
 condition, 69, 107
 high pass, 169
 selection of, 71
filtration, 69
 off-line, 108
fittings, fluid resistance in, 191

flow
 coefficient, 117
 variation, 118
 control, 53, 83, 85, 87, 219
  variable, 87
 force, 120, 121, 193
 leakage of, 111
  internal, 17
 motor, 21, 186
 orifice, 117
 pipes, in, 113
 pump, 59, 95, 98, 198-9, 202, 211
 turbulent, 111
fluid(s)
 bulk modulus, 137
 cleanliness level, 67
 compressibility, 36, 138, 143, 145, 150
 dynamic viscosity, 129
 momentum forces, 112
 resistance, 191-2
 shear, 130
 type of, 9, 11, 33
 momentum, 119
fork lift truck, 80
 systems, 30
four way valve(s), 79, 80, 87
frequency,
 hydraulic natural, 150
 response,
  151, 155, 164, 166
  analysis, 169
  tests, 169
friction factor, 114-5, 188, 192

G

gantry crane, hydraulically powered, 183
gas laws, universal, 66
gear pump(s), 9, 11, 13, 15, 19
 external, 11
Gerotor, 22
Gerotor principle, 20

H

heat generation, 16, 63, 72, 213

Index

high pass filter, 169
high-pressure filter, 107
HSLT motors, 24, 25
hydraulic
   actuators, 30
   motor(s), 10, 37, 102, 181, 184, 223
   capacity, 216
   natural frequency, 150, 155-6, 165, 172
   potentiometer, 99, 221
   power, 3, 137, 215, 217
   presses, control of, 237
   pumps, 53
   stiffness, 143-4, 171
   system,
      management of, 233
      closed loop, 144
hydromechanical servo, 98
hydrostatic
   motors, 9, 10, 20, 215
   pumps, 9
   system, 171
      rotary, 171
   transmission circuit,
      actuator, 104
      rotary, 103
   transmissions, 19, 78, 102
hysteresis, valve, 160

I

injection moulding machine, 207
input
   demand signal, 140
   signal(s), 141, 166, 170
      amplitude, 170
      range of, 142
integrator, 140
internal gear pumps, 12, 13

K

kinematic viscosity, 112, 113

L

laminar flow, 111, 114, 116, 129

leakage losses in pumps, 127
leakage range, 158
linear actuator(s), 44, 30, 83, 138
hydrostatic transmission circuit, 104
transmissions, 103
load(s)
   control
      circuit, 105
      valves, 46, 105
   mass,
      effect of, 137
   inertial, 145
   regenerative, 104
   sensing, 16, 77, 97-8
      control, 99
      system, 224
low speed motors, 20
LSHT motors, 24, 25
lubricated slippers, 15

M

magnetic hysteresis, 160
maintenance, 25, 233
   predictive, 235
   procedure, type of, 234
   procedures, 67
Mean Time Between Failures (MTBF), 238
mechanical
   efficiency, 17-8, 127, 131-2, 216
   hysteresis, 156
   loss, 17, 127, 130-1, 133, 185
meter-in, 56, 209
   circuit, 83
   control, 83, 84, 86-7, 175
meter-out, 56, 209, 211, 212
   control, 85-6, 94, 97, 105, 211-2
metering, 202
   asymmetrical, 95
   characteristics, 97
   edges, notched, 97
   non-symmetrical, 94
   notches, 94

restrictive, 59, 102
momentum force, 119
monitoring, 239
   signals, 55
Moody Chart, 114-5
motor,
   axial piston, 15
   brake circuit, 103
   controlled systems, 104
   creep, 23
   displacement, 20-1, 104, 181-2, 186, 208
      control, 103
   drain, 23
   efficiencies, 202
   flow, 21, 186
   hydraulic, 102, 181, 184
   hydrostatic, 215
   low speed, 185
   Orbit type, 22
   performance characteristics, 18
      comparison of, 23
   radial piston
      cam type, 21
      eccentric, 21
mountings, actuator, 31
multiple actuator circuit, 86

N

Natural frequency, 205
Nichols Chart, 169
non-symmetrical valve
   metering, 94
notched metering edges, 97
notches, 97
null position, 156

O

oil
   cooling, 211
   viscosity, 181
      variation, 113
open
   centre valves, 81, 95
   loop
      circuits, 16

frequency
response, 206
  gain, 155
  performance, 140
  system, 138, 151
  tests, 170
  time response, 140
  transfer function, 153, 156, 168
  velocity control, 138
Orbit motors, 22
orifice equation, 93
orifice flow, 117

P

parallel leakage spaces, 116
particle sizes, 69
particles, 71
  contaminant, 63
Pascal's Law, 111
perturbation
  analysis, 159
  technique, 157
phase advance
  characteristic, 166
  compensator, 166
  frequency response, 167
PID (see proportional, integral and derivative con), 167
  system, 168
Pilot Operated Check Valve, 82
pilot operated check valve (POCV), 82, 106
  circuits, 105
pilot operated valves, 106
pipe,
  flow through a, 123
  pressure losses in, 112, 190
piston pumps/motors, 14-5
plate valve, 15
Poise (P), 112
polytropic expansion index, 65
poppet valve(s), 42, 45, 119

position
  control loop, 168
  system, 161
positive displacement pumps, 11
potentiometer, hydraulic, 99, 221
power
  control,
    constant, 100
    pump, 220
  efficiency, 87
  hydraulic, 137, 217
  pump
    input, 217
    output, 220
  transmission efficiencies, 170-1
predictive maintenance, 235
pressure and torque relationship, 17
pressure compensated bypass valve, 98
pressure
  compensated
    flow control valve (PCFCV), 83
    pump, 57, 87
    valve(s), 56-8
  compensation, 16, 97, 219
  compensator, 219
  control, 86, 162, 163
    valve, 41-2, 111
  feedback, 48, 99, 168-9
  /flow characteristics of restrictors, 112
  gain, valve, 158
  loss in
    filters, 107
    pump inlet, 113
    pipe, 112, 188, 213
    valve, 212
  losses, 72, 115, 190, 212
  piping, for, 111
  reducing valve, 44
  relief valve, 57
  shock control, 151
primary control, 102

proportional
  compensation, 164
  control, 165-6
    valve, 55
  integral and derivative (PID) contro, 167
  solenoid, 88
  valve(s), 209, 211
pulsation absorption, 65
pump,
  boost, 102
  control(s), 78, 81, 218, 222
    strategy, 219
    variable displacement, 98
  controlled systems, 102, 170
  displacement, 16
  controller, 171
  controls, 16
  mechanism, 99
  efficiencies, 202
  efficiency, 127, 215
  external gear, 12-3
  gear, 9, 11, 15, 19
  inlet pressure, 192
  input power, 217
  internal gear, 12-3
  leakage, 100
    losses in, 127
  operating characteristics, 100
  output power, 220
  performance
    characteristics, 18
    of, 127
  piston, 9, 111
  power, 219
    control, 220
  pressure compensated, 57, 87, 104
  radial piston, 15
  unloader valve, 222
  unloading circuit, 105
  vane, 9, 11, 13-4
  variable
    capacity, 215
    diaplacement, 53, 79, 86, 171, 219

Index

volumetric efficiency, 218
purge valve, 102

R

radial piston
　cam type motor, 21
　eccentric motor, 21
　motors, 20, 24
　pump, 15
ramp time, 210
reducing valve, 45, 103
regeneration, 16
reliability, 3, 34, 233, 234-5, 238
　level of, 234
　system, 69, 234, 238
relief valve(s), 41-2, 81, 83, 87, 96, 98, 102, 107, 111, 175-6, 200, 215, 222, 236
　flow, 213
　poppet type of, 193
　pressure, 211
　setting, 216
　single stage, 42, 193
　two stage cartridge, 44
　　cross line, 189-90
　　dual, 44
reservoir design, 74
reservoirs, 64, 72
resistance, fluid, 192
restrictive
　control, 87
　metering, 59, 102
restrictor(s), 34, 53, 58, 85, 176
　adjustable, 36
　control, 79
　pressure drop, 177
　pressure/flow characteristics of, 112
　size, 195
　valve, 56, 83, 176
　　adjustable, 176, 194
rotary
　actuator(s), 37, 83
　　performance, 38
　vane, 38
hydrostatic
　systems, 171
　transmission circuit, 103

S

safety, 233, 238
　interlocks, 237
secondary control, 104
sensing system, load, 224
sensitivity, contaminant, 19
sequence valve, 106
servo,
　hydromechanical, 98
　system(s), 95, 99
　valve control, 172
　valves, 168
servovalve, electrohydraulic, 55
sharp edge orifice, 118
shear stress, viscous, 203
shock alleviation, 65
side
　loading, 10
　loads, 38
　plates, 12
single
　ended actuator, 177
　stage relief valve, 42, 193
spool, 42, 59
　type valves, 54
　valve(s), 45, 56-7, 82, 99, 111, 117, 121-2, 196, 202
stability criteria, 155
steady state
　accuracy, 156, 159, 161, 225
　error, 164-5, 169, 229
　gain, 164, 166
　operating condition, 149
　performance, 137, 159, 220
step response, 143
stiffness,
　hydraulic, 143-4, 171
mechanical, 162, 164, 170
suction filters, 107
swash plate, 15, 24
units, 15
symmetrical valves, 94
system
　cleanliness, 235
　duty cycle, 72
　dynamic performance, 170
　efficiency, 185
　frequency response tests, 169
　gain, 205
　performance, 138

T

tandem centre valves, 81
thermodynamic aspects, 73
three-way valves, 80
throttling action, 45
tie rods, 30
time constant, 137
time response, 142
torque
　control, pump, 220
　pressure relationship, and, 17
　ripple, 23
track drive, 216-7, 220
transfer function, 137, 226
transmission, 216
　circuit,
　　linear actuator hydrostatic, 19, 78, 102, 104
　　rotary hydrostatic, 103
　linear actuator, 103
trunnion mounting, 31
turbulent flow, 111
two
　position valves, 79
　stage cartridge relief valve, 44

U

underlap, valve, 157

underlapped valve, 159
underlapping, 158
unloading valve, 101, 106
unreliability, 63
upstream pressure, 121
upstream pressure force, 120
upstream velocity, 120
utilisation, 34

V

valve(s)
    actuator
        circuit, 138-9
        control, 138
        dynamic response, 144
        system, 138, 141, 151, 153, 237
    /actuator system, 204
    brake, 190
    bypass, 79
    central bypass, 59, 96
    characteristics, 47, 139, 145, 157, 159-60, 199
    check, 82, 97, 102
    circuits, pilot operated, 105
    closed centre, 81, 97
    control, 55-6, 72, 170-1, 177, 219
        servo, 172
    controlled systems, 34
    counterbalance, 46, 105, 189
    cross line relief, 189-90
    cushioning, 38
    directional control (DCV), 54, 224, 237
    dynamic performance, 155
    electrohydraulic, 141, 150
    flow
        characteristics, 88
        control, 53
        gain, 146, 155
    force analysis, 118
    four-way, 79, 87
    gain, 141, 155

hysteresis, 160
leakage, 157, 225
load
    check, 96
    control, 46, 105
    holding, 82
metering
    area, 96
    ratio, 95
open centre, 81, 95
performance characteristics, 202
pilot operated, 106
poppet, 119
pressure
    compensated, 56
    control, 41, 111
    drop, 93, 142
    flow characteristic, 121
    gain, 158
    loss, 212
    reducing, 44
proportional, 209, 211
    control, 55
pump unloader, 222
purge, 102
rated flow, 177
reducing, 45, 103
relief, 41-2, 102, 111
restrictor, 56, 83, 176
single stage relief, 42, 193
servo, 168
sizing, 92
spool, 196, 202
    type, 54
systems, counterbalance, 48
tandem centre, 81
three position, 80
two
    position, 79
    stage cartridge
relief, 44
underlap, 157
underlapped, 159
unloading, 101, 106
zero lap, 158
vane, 19, 23, 37

motors, 13
pumps, 9, 11, 13-4
rotary actuator, 38
variable
    capacity pump, 215
    control, 56
    displacement pump(s), 79, 86, 171, 219
        control, 98
        pressure, 99
vehicle
    drive(s), 16, 25
    transmission, 215
velocity control, 77, 83, 102, 161
    open loop, 138
vena contracta, 117
vibration, 107, 151, 238
viscosity,
    dynamic, 112, 203
    kinematic, 112, 113
    variation, oil, 113
    force, 116
    friction, 130, 163
voltage
    error signal, position, 171
    signal, 141
volume, accumulator, 209
volumetric
    efficiency, 17-8, 72, 111, 127, 131-2, 188, 209
    flow loss, 128-9
volumetric losses, 127, 133

W

water coolers, 72
weight loaded system, 196
wheel
    drive, 184-5, 190
    torque, 185
winch, 181
    application, 179
    drive, 16
winches, 20, 103, 105

Z

zero lap valve, 158